Thinking Through Science

Activities from Physics 1

Toni Smith   Carol Tear   Alan Yate

SERIES EDITOR: DAVID EDWARDS

UNWIN

HYMAN

Published in 1989 by
Unwin Hyman Limited
15–17 Broadwick Street
London W1V 1FP

British Library Cataloguing in Publication Data

Smith, Toni
    Activities from physics.
    Bk. 1
    I. Title II. Tear, Carol III. Yate, Alan
    530

ISBN 0–7135–2847–8

Typeset by MS Filmsetting Ltd, Frome
Origination by Chroma Graphics, Singapore
Printed and bound by New Interlitho, Italy

Cover photograph by Martyn Chillmaid
Book design by Juliet and Charles Snape

Contents

About this book

These exercises are about important applications of physics. Some of the information you will have to read—some of it you will have to get from the tables, charts and pictures. You will use this information and what you already know about physics to answer the questions.

Thinking Through Science, Physics 1 will help you to understand some useful ideas in physics. But most of all, we hope that the book will let you enjoy physics—a very important part of our everyday lives.

1 MARATHON

40 000 feet hit the pavements and roads in the London Marathon. Good footwear could mean the difference between success and failure...

upper · heel tab · midsole · outsole

When a runner's foot strikes the ground, the movement energy has to be absorbed without damaging the runner.

- Six runners agreed to test different training shoes.

- A problem with many trainers is that the makers have made a feature of the heel tab. In some trainers, this tab rubs against the Achilles tendon and could cause injury.

- There is a lot of information about the different shoes here. It is difficult to compare the shoes with information in this form—a table would help.

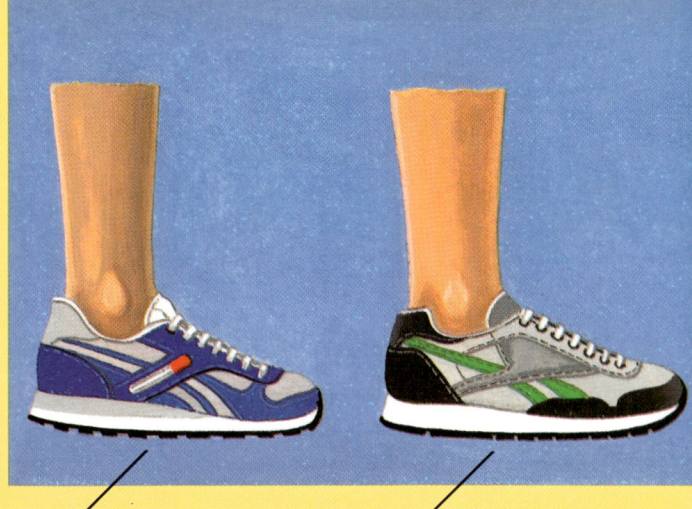

Zoom: £28.
A special hard-foam midsole. Tested over 350 miles.

"The midsole was very hard, but made a good shock absorber. They had good grip in the dry, but not on wet surfaces. I had to cut off the heel tab because it aggravated my Achilles tendon."

Cheetah: £19.99.
Double-layer midsole. Tested over 600 miles.

"The shock absorbancy was good. I found these shoes very comfortable, the cut-away heel gave no problems. The grip was not so good in wet or icy conditions. I would certainly buy these."

1. Construct a table with the following headings and fill in the information on all the shoes:

   Name   Grip

   Price   Comfort

   Distance tested over   Any other information

   Midsole construction

2. Which shoes were the cheapest?

3. Which were tested over the longest distance?

4. How many different types of midsole construction were there?

5. The tests on the shoes did not give a fair comparison. What THREE things would need to be kept the same each time the test was carried out in order to make it a fair comparison?

The soles of running shoes are made of **elastic materials.** When elastic materials are compressed they absorb the movement energy. When the force is removed, they spring back into shape.

- **Plastic materials** can be compressed permanently. They don't spring back into shape.

- Look at these three materials being tested with a 50 kg (about 8 stone) mass.

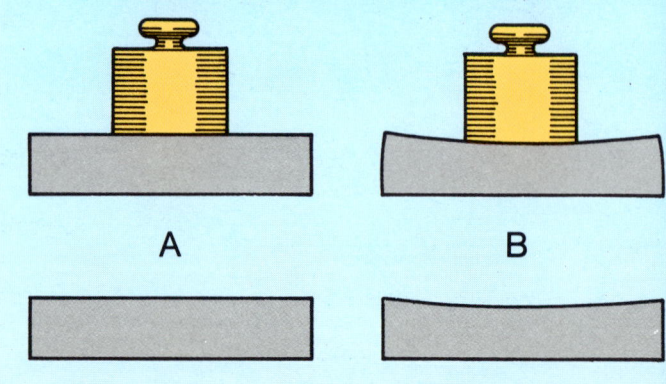

A   B

Laser: £42.
Three-layer midsole.
Tested over 450 miles.

"The midsole started to squash and lose its shock-absorbing properties. Good grip, comfortable to wear. The heel tab was no problem."

Sprinter: £38.50.
Special hard-foam midsole.
Tested over 400 miles.

"The shock absorbancy was fine. The outsole is perfect, no trouble with grip. The heel was too wide—I had to wear thicker socks, but I'd recommend these shoes."

Speedo: £49.95.
Air-layer sandwiched in foam in the midsole.
Tested over 900 miles.

"I felt as though I was floating. They have a good grip on wet surfaces. The heel tab caused me no problems."

Apollo: £26.99.
Three-layer midsole.
Tested over 400 miles.

"The midsole is firmer than my usual shoes, but I found it still absorbed the shock OK. The outsole wore down rather quickly. Grip was good on roads, and not too bad on moorland. Very comfortable."

6. Which of the materials **A**, **B** or **C**, would you use to make the midsoles of training shoes?

7. Which of the materials shows plastic behaviour?

8. Which of the materials is most difficult to compress?

9. Sport provides many examples of the use of elastic materials. Describe one use of elastic materials in:

   (a) tennis

   (b) bicycle moto-cross (BMX)

   (c) hockey

   (d) athletics

10. A long jump pit uses a **plastic** material. What is it? Why do long jumpers need to use a plastic material?

ARTIFICIAL LIMBS

Advance in artificial leg technology!

British technologists have developed a new system which promises to reduce waiting lists for replacement limbs.

    Researchers at University College, London have produced a computer-controlled system to aid the design and fitting of limbs. The hope is that patients will visit the centre and walk away on their new leg the same day.

There has been little progress in this field over the past 40 years, and there is more demand for replacement lower legs than other limbs. This has meant a long wait for people like Mr Joe Hinks. Mr Hinks had an industrial accident which led to his leg being amputated below the knee.

    The new system will use light to take accurate measurements of the patient's stump. A special computer program then designs the best shape for the top of the new limb so that it fits the stump perfectly.

    The material used to make the artificial limb has to be chosen carefully. It must be light so that the limb is not a burden to the person, and yet must not stretch or get squashed under pressure. Not many materials fit these requirements and a lot of research had to be done to find the ideal one.

The fairest way of comparing the 'lightness' of materials is to find their densities. The **density** of a material is the mass in grams of one cubic centimetre of it. The units for density are g/cm³.

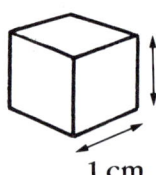

1 cm

The table shows the densities of some of the materials that researchers thought would be suitable for making artificial limbs.

Material	Density g/cm³
Acrylic	1.2
Aluminium	2.7
Nuvo-plastic	0.5
Polypropylene	0.9

1. Why do you think there is a long waiting list for replacement lower limbs?

2. Two properties needed in the materials used for making artificial limbs are mentioned in the article. What are they?

3. Suggest THREE other properties needed in the material.

4. Why is using density measurements a fair way of comparing the lightness of materials?

5. If density were the only property that mattered, which material in the table would be used for making artificial limbs?

6. A typical artificial leg has a volume of 1000 cm³. What mass of the material you chose in question 5 would be needed to make this leg?

These two sets of apparatus are used to find the tensile and compressive strengths of materials. The tensile strength of a material tells us the mass it can hold before it stretches and breaks. The compressive strength tells us the mass that can be supported before it is crushed.

This table shows the results of these tests.

Material	Tensile strength (maximum load before stretching)	Compressive strength (maximum load before crushing)
Acrylic	1.5 kg	1.3 kg
Aluminium	0.8 kg	0.85 kg
Nuvo-plastic	0.08 kg	0.3 kg
Polypropylene	0.35 kg	1.1 kg

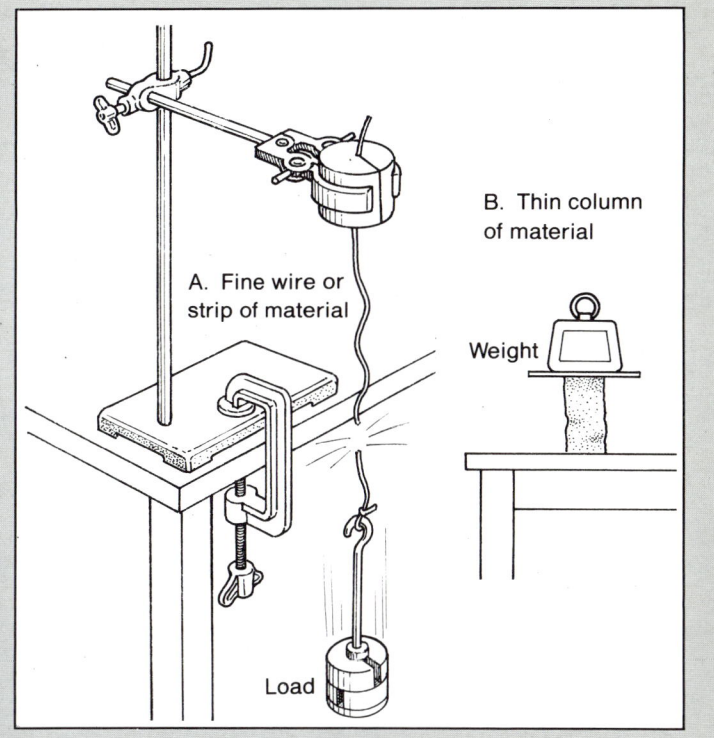

A. Fine wire or strip of material

B. Thin column of material

Weight

Load

7. Explain the difference between tensile and compressive strengths.

8. Which set of apparatus, A or B, tests for tensile strength?

9. Suggest TWO things that must be kept the same when comparing the compressive strength of different materials.

10. Which is the most important property of the material to be used for replacement limbs—tensile strength or compressive strength?

11. If strength were the only property that mattered, which material in the table would be used for making artificial limbs?

Some of the other properties of the materials.

Material	Hardness*	How easy it is to engineer	Cost
Acrylic	20	FAIR	MEDIUM
Aluminium	24	HARD	MEDIUM
Nuvo-plastic	12	EASY	HIGH
Polypropylene	18	EASY	LOW

*The harder the substance, the higher the hardness number.

You are going to use your judgement as a scientist to advise the technologists of the best material from which to make the artificial limbs. These are the properties you should consider:

(a) Density   (b) Strength   (c) Cost
(d) Hardness   (e) Ease of engineering

12. Draw up a table like this. Fill in the materials in alphabetical order. Write in the five properties in order of importance, with the most important first in the table.

PROPERTY MATERIAL					TOTAL SCORE

13. Now give each material a score for how suitable it is. For example, if when considering density you think Nuvo-plastic is very suitable, give it a high score, like 5:

Very Suitable 5 4 3 2 1 0 Not Suitable

You should end up with five marks for each material. Add these up to give a total. The highest score will show the most suitable material.

14. Write a brief report recommending a material to use for artificial limbs. You should mention:

(a) the properties you used to judge each material,
(b) why you thought some properties more important than others.

UP, UP AND AWAY

It was a perfect morning for our journey. There wasn't a cloud in the sky, and no wind at all on the take-off field.

We turned on the burner and began to inflate the balloon. At first the bag was spread out on the ground, but gradually it filled up with hot gas and rose into the air. The mooring ropes tightened as the balloon tried to float away. We quickly climbed into the gondola and the ground crew set the balloon free.

The burner was still on, and the balloon rose steadily higher and higher. It began to get colder. We couldn't feel any wind because a balloon moves with the wind. There must have been a strong breeze because we were moving quite fast over the roads and houses below. They looked tiny.

Now we were high enough, and to keep the balloon at a steady height we kept turning the burner valve on for five seconds and off for twenty seconds. When the burner was off it was very quiet. The wind was blowing towards the south, and we travelled with it over villages, rivers and the railway track.

We had travelled a long way when clouds began forming around us. As we entered the first cloud, it suddenly got colder and the balloon began to rise. We wondered whether to try to get above the cloud, but we didn't know how far up it went, and we had no oxygen. We pulled the white vent line to let out some of the hot air so that we would fall to below the cloud. We closed it quickly, so that we would not fall too far. As we came out of the cloud, we saw the houses and people, closer now, and realised we had better look for somewhere to land safely.

We could not come down in the town, so we turned on the burner valve. We kept the balloon at a safe height by switching the burner valve on and off.

Luckily, we had passed over the town when the supply of gas was used up and the balloon began to descend again. It was getting very low just when we came to a reservoir! It looked as if we would be landing in the middle of it. Quickly we grabbed a couple of the sandbags and threw them out. This made the balloon lighter, so it began to rise again.

We watched the countryside carefully, and when we had cleared the trees and found ourselves crossing open fields, we pulled the white vent line again to let out some of the hot air. The balloon began to descend, and after a while the gondola almost touched the ground, but the bag was still in the air and the wind began to blow it across the field towards some trees! Quickly we broke the safety ties and tugged on the red rip line. This pulled the panel out of the top of the balloon and released all the hot air. The gondola stopped with a bump as the bag collapsed on the ground. We had landed.

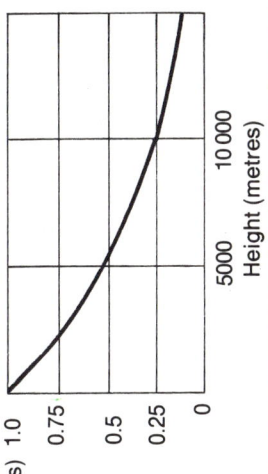

This graph shows the air pressure at different heights above the earth.

1. Why couldn't the crew feel the wind?

2. Name the force of attraction which pulls the balloon to the ground.

3. What causes the force pushing the balloon upwards?

4. What can you say about the value of the pushing and pulling forces on the balloon when it floats at a steady height?

5. Explain how the crew could keep the balloon at a steady height.

6. What are the THREE ways of making the balloon fall mentioned in the text?

7. Why would the crew need oxygen if they had gone up higher?

8. What happens to the value of the air pressure as you rise above the Earth?

9. Copy and complete this key with the appropriate letters and description from the chart.

Letter	What happened at this point in the flight
A	'the ground crew set the balloon free'
	'as we entered the first cloud'
	'we grabbed a couple of sandbags and threw them out'
J	'we pulled the white vent line again'

10. What was the air pressure at point B?

11. How many hours after leaving the ground did the balloon reach point G?

12. (a) Which letter marks the highest point reached by the balloon?

    (b) What was the air pressure at this point?

    (c) Use the graph of air pressure against height to find the height of the balloon at this point.

During the flight an instrument recorded the air pressure. The pen moved steadily across the paper. As the pressure went down, the pen moved down. This is the chart that was produced during the flight:

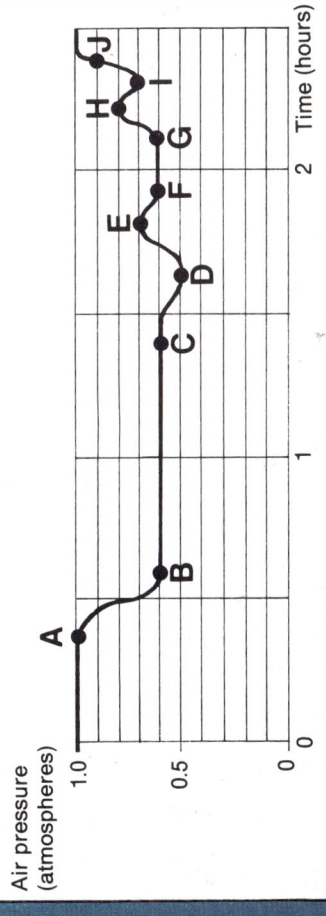

4 KETTLE

Our market research shows that there is a demand for a picnic kettle which could run off a 12 volt car battery.

It should be able to boil ½ litre of water in under 4 minutes. The kettle must be safe and easy to handle.

Can you supply us with a suitable design? We will work out the costings for production and marketing.

This is Sue, the chief technician in the Research Department.

The first thing she does when she receives the memo is to look in manufacturers' catalogues to see if there is a suitable heating element for the picnic kettle.

Sue finds that there are two heating elements, but which one would be best?

1. Give one advantage and one disadvantage of using a car battery as the source of energy for a picnic kettle.

2. Why should the coils be covered with a material which is a thermal conductor, but an electrical insulator?

The Element Company Catalogue
Heating elements – same power rating

(1) **The Flat Coil**
Nichrome wire coiled tightly to save space. Enclosed in a specially toughened casing for safety. Available in two sizes.

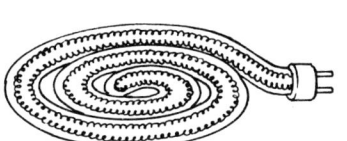

(2) **The Multi-Use Coil**
Nichrome wire coiled in a vertical tube. Enclosed in a non-conducting material for safety.

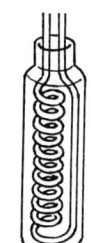

Sue decided to do some experiments to find out which, if either, of the heating elements would be suitable for the picnic kettle.

She set up an electrical circuit with a normal 12-volt car battery to heat some water. She measured the temperature rise of the water for each heating element. Here are her results.

HEATING ELEMENT 1	Temp °C	4	20	36.5	52.5	62	85					
The flat coil	Time in secs	0	30	60	90	120	150					
HEATING ELEMENT 2	Temp °C	5	13	21	31	37	45	53	61	69	77	85
The multi-use coil	Time in secs	0	30	60	90	120	150	180	210	240	300	360

3. Sue took care to make the experiments a fair comparison of the heating elements. Suggest THREE things that needed to be kept the same in each experiment.

4. Plot Sue's results on graph paper. Label the vertical axis: Temperature from 0°C to 100°C. Use the same graph for both sets of results.

5. Use your graph to decide if any mistakes were made in measuring the temperatures during the experiments.

6. Estimate the time taken for each heating element to boil the water.

The Design Department came up with three designs for picnic kettles. They all used the flat coil heating element which Sue recommended.

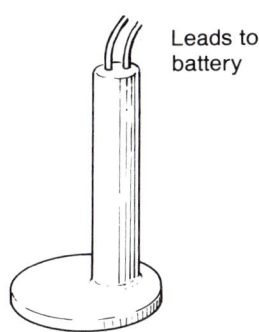

Leads to battery

SPARTAN STYLE

Coiled (flat coil) element encased in plastic
For use in any vessel
Only 10 cm tall

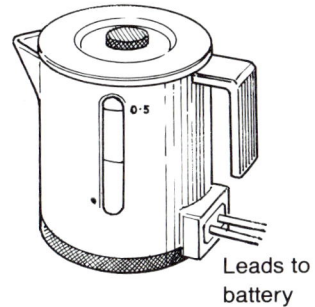

MODERN STYLE

Many colours available
Visible fill level
Strong plastic insulated body
Capacity 0.5 litres

Leads to battery

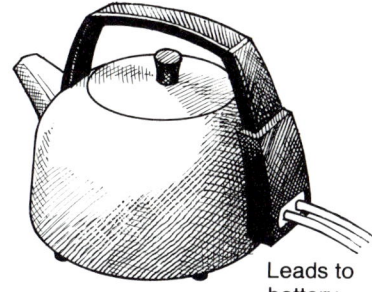

TRADITIONAL STYLE

Shiny metal case
Flat heating element

Leads to battery

Sue had to test each of the designs to find out if they were safe and if their performance matched the requirements.

She boiled half a litre of water in each kettle. Here are her results.

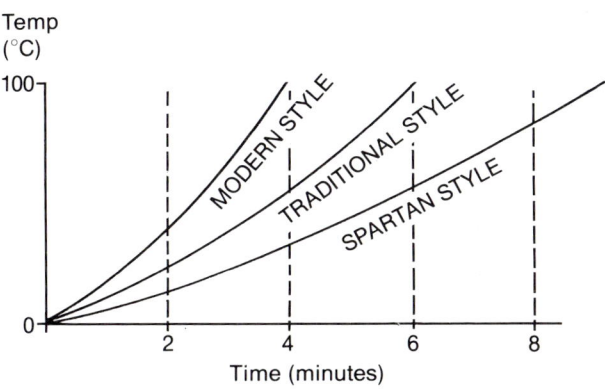

7. Estimate how long each kettle takes to boil the water.

8. Suggest why the times are not the same as those in Sue's first test of the flat coil.

9. Which of the kettles will NOT be chosen? Explain why.

10. Why are the heating elements placed at the bottom of the kettles?

11. The modern and the traditional styles have the same heating element. Suggest why they take different times to boil the same amount of water.

12. Why might the kettles take longer to boil half a litre of water when they are being used on a picnic?

13. Suggest TWO things that should be checked to make sure the designs are safe.

14. The costs of making and selling the kettle can be grouped under different headings. One would be 'Cost of materials and parts'. Suggest THREE other types of costs involved.

15. Use all the information you have examined to write a summary report to the Marketing Department. Use a memo like the one shown at the start. Your report should be about 500 words in length and include:

   ● details of the different types of kettle,

   ● the tests of the heating elements,

   ● the test of the three designs,

   ● your recommendation on which kettle to manufacture.

5
ON THE MOVE

1. Match these speeds to the five moving objects in the drawing:

   (a) 1 m/s   (d) 600 km/h
   (b) 10 km/h   (e) 50 miles/h
   (c) 4 km/h

2. The units for speed have two parts. One part gives a distance. What is the other part?

3. Why do we use different units for measuring the speeds of different things?

4. Which of the units for the speeds in question 1 is the odd one out? Why?

5. Give TWO reasons why it may be important to know the exact speed of a vehicle.

When dealing with speeds, we use the **SPEED EQUATION**. This can be written in three different ways.

1. SPEED = DISTANCE ÷ TIME	2. DISTANCE = SPEED × TIME	3. TIME = DISTANCE ÷ SPEED
We use this if we need to calculate a speed.	We use this if we want to find out how far we can go in a certain time.	We use this if we need to find the time it takes to travel a certain distance.

6. Copy each of these equations into your book, explaining when to use each one.

Here are two cars.

The mini can travel from London to Liverpool, 300 km, in 5 hours.

The sports car can go from Bath to Nottingham, 100 km, in just 1 hour.

7. What are the average speeds for the two cars on their journeys?

8. How long would it take the sports car to travel from York to London (450 km) at the same average speed?

9. If the mini took $2\frac{1}{2}$ hours to travel from Dundee to Edinburgh at the same average speed, what is the distance between the two cities?

This Lotus is typical of modern Formula 1 racing cars. It has a maximum speed of 200mph. Its bodywork is made of a fireproof, light-weight material.

10. Why is the bodywork made of a fireproof *and* light-weight material?
11. The circuit of a Grand Prix track is 2 miles. In a Formula 1 race, the Lotus covers the 60 laps of the race in 45 minutes. What is the average speed of the car during the race?

The space shuttle is carried to the launch pad on a transporter. Transporters are the world's largest trucks and crawl along the roadway at a snail's pace—1.5 km per hour.

12. Why does the transporter move so slowly?
13. How long would it take the transporter to move the 12 km to the launch pad?

The Blackbird spy plane is made of a titanium alloy which withstands the very high temperatures caused by air friction. In 1976 an SR-71 set the world speed record at 3529 km per hour.

14. Why must the Blackbird have a smooth, streamlined shape?
15. The Blackbird can travel over the length of Yugoslavia in 1 hour. If the length of Yugoslavia is 3180 km, what is the average speed of the plane in kilometres/minute?

Staffield County Council are trying to decide where to site their new Fire Station.

A new government directive says that fire engines MUST be able to reach 'risk areas' in under 3 minutes.

Possible sites are marked A and B.

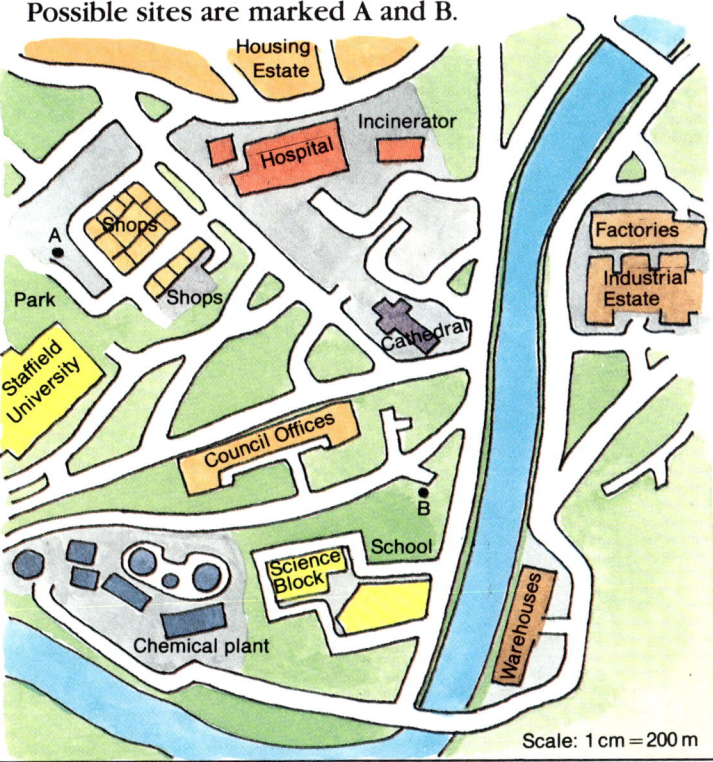

Scale: 1 cm = 200 m

16. Which are the THREE major risk areas in Staffield?

17. Copy and complete this table. Use the edge of a piece of paper to measure the distances from the sites to the three risk areas.

NAME OF RISK AREA	DISTANCE TO SITE A (m)	JOURNEY TIME (s)	DISTANCE TO SITE B (m)	JOURNEY TIME (s)
1.				
2.				
3.				

18. The maximum speed of an engine is 15 m/s. Calculate how long it takes to get to EACH risk area from site A and from site B.

19. Use your answers to questions 16, 17 and 18 to help you write a letter reporting your investigation to the County Council. Your letter should include:

- the background to the problem
- the risk areas you have identified
- the measurements and calculations you made
- the site you recommend for the new fire station

6

BRAKE TEST

The time it takes to stop a car depends on its speed. The car continues to travel along the road when the driver puts the brakes on. The faster the car is travelling before braking, the greater the distance before it stops.

1. What is the stopping distance (the 'thinking distance' plus the 'braking distance') of the car if it is travelling at
   (a) 30 mph,   (b) 55 mph?

2. What happens to the stopping distances as the speed of the car increases?

3. Suggest what is meant by the 'thinking distance'.

4. Write down THREE things that could affect a driver's 'thinking distance'.

5. Why does the 'thinking distance' increase as the speed increases?

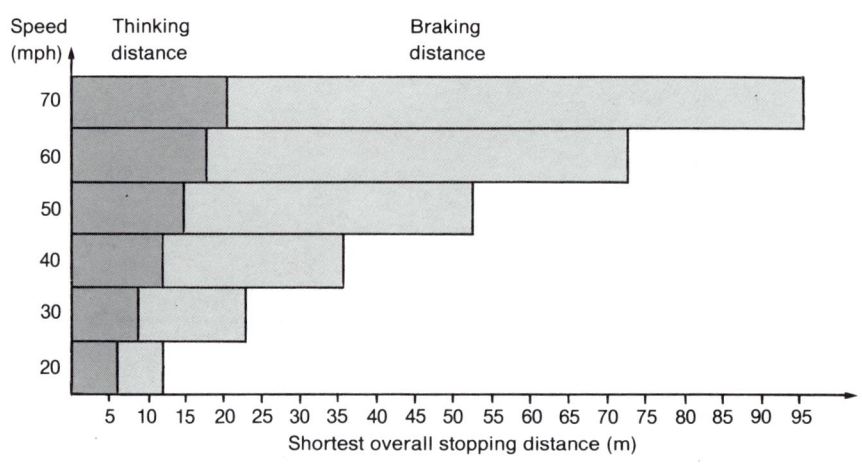

Speed (mph) — Thinking distance / Braking distance — Shortest overall stopping distance (m)

Brakes work by stopping the wheels turning. One type of brake is called a disc brake.

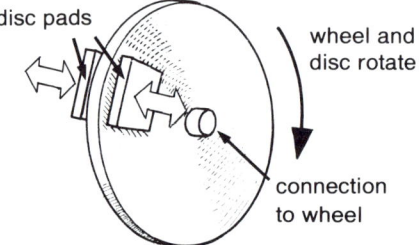

disc pads

wheel and disc rotate

connection to wheel

movement of pads: pads press on the disc to stop rotation

The pads must be rough and dry.

6. What is the force which stops the disc from going round called?

7. The car's movement energy is not lost. What form of energy is it changed into?

8. What would happen if
   (a) grease got on to the disc?
   (b) the pads wore away and the surfaces became smooth?
   (c) the pads and disc were wet?

When a driver suddenly slams on the brakes, the wheels of a car stop turning in a very short time. But the car is heavy and has energy of motion. It carries on moving and the wheels skid along the road before the car comes to a stop. For the car to stop in the shortest distance possible, the wheels must not skid. Water on the road will increase the risk of skidding.

The tread pattern on a tyre helps to clear away water from between the tyre and the road.

9. Suggest FOUR things which would make a car more likely to skid.

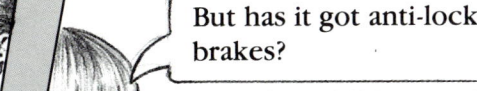

I'm looking for a car which is about three years old, with low mileage, and it must have anti-lock brakes.

Have I got just the car for you! Look at this ... 2 years old, 30 000 miles on the clock—what a bargain!

But has it got anti-lock brakes?

Are you sure you want them? You should read this report. Come back when you've looked at it, but don't wait too long—the car's a bargain!

ANTI-LOCK BRAKES: TEST REPORT

Some cars are now fitted with anti-lock braking systems. If you slam your foot on the brake, you can stop the car wheels turning completely, but this can mean that the car will skid along the road, taking longer to stop. To stop in the shortest possible distance means braking without skidding. Normally, if you start to skid you must take your foot off the brake and then brake again - difficult to do in an emergency. Anti-lock brakes do this for you. They have sensors which detect when the car is going to skid and disconnect the brakes for a fraction of a second. Anti-lock brakes could be a great improvement. But do they work properly?

We asked two drivers to bring their cars along to our test centre to try them out.

Mr Williams drove his four-door family car around our test circuit. This car has anti-lock brakes. When braking from 50mph to a stop, the overall stopping distance was 53 metres.

Two days later, Ms Foster drove her sports car around the same circuit. This car does not have anti-lock brakes. When braking from 30mph to a stop, the overall stopping distance was 23 metres.

Our conclusion from this test is that anti-lock brakes are not an improvement on the driver's own skill. We advise our customers not to buy cars with anti-lock brakes.

10. Write down FIVE reasons why the test was not a fair test of the anti-lock brakes.

11. Write out how you would do the test. Use these questions as a guide:

- What hypothesis are you testing? (What suggestion are you testing?)
- What equipment will you need?
- What will you keep the same each time you do the test?
- What will you change each time?
- What will you measure?
- How will you check your results?
- How will you know if the hypothesis is correct?

Hello again. Did you read that report? Those brakes are a waste of money aren't they?

So they said, but I wasn't convinced. It wasn't a fair test.

Oh really?

Well for a start they used different drivers ... and cars! ... and it doesn't say anything about the tyres, and ...

So you won't want that car then.

Never mind. As it happens I've got just what you're looking for down here. It's a 2 year old ...

7
WORKING UNDER PRESSURE

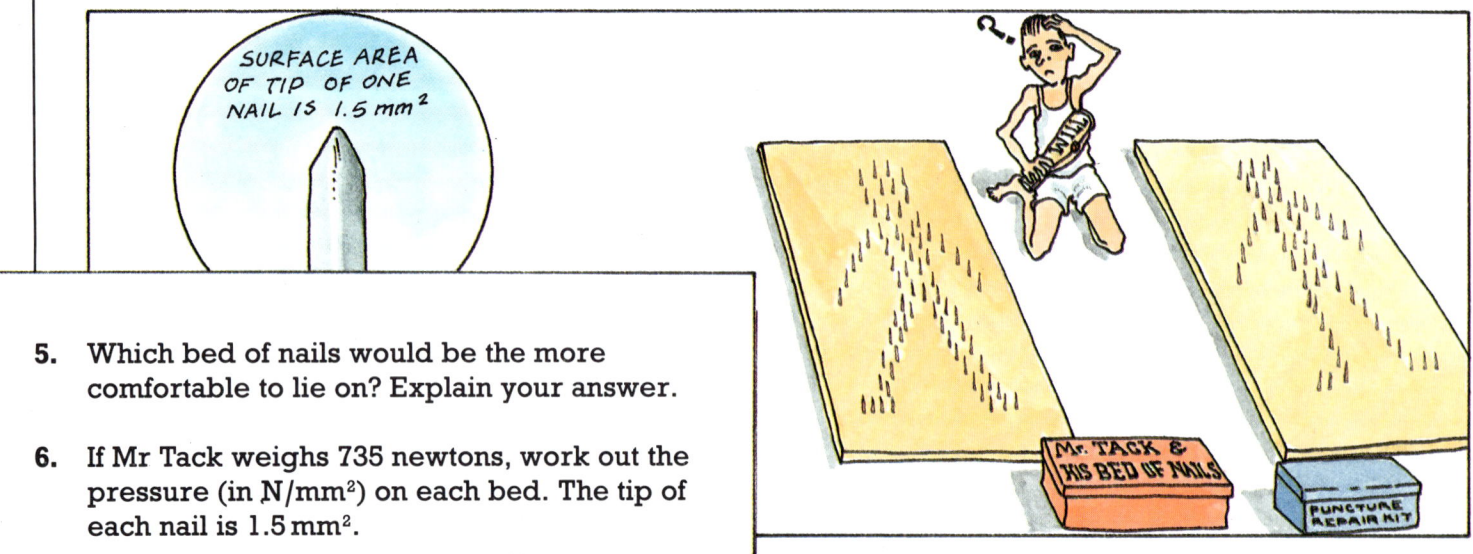

1. Why has wheelbarrow 1 sunk into the ground on the site, when wheelbarrow 2 seems all right?

2. The man with wheelbarrow 1 could do something about his load to stop it sinking into the ground. What should he do?

$$PRESSURE = \frac{FORCE}{AREA}$$

3. Wheelbarrow 1 could still move the same load of bricks if the worker used something from the site to help him to get over the mud. What should he use? How will this help him?

4. Why do the bases of the scaffolding poles have flat metal plates fitted to them?

5. Which bed of nails would be the more comfortable to lie on? Explain your answer.

6. If Mr Tack weighs 735 newtons, work out the pressure (in N/mm^2) on each bed. The tip of each nail is $1.5\ mm^2$.

7. Why is the bag with the thin strap more uncomfortable to wear?

8. Why do the stiletto heels keep getting stuck in the ground?

9. What could be done to prevent making the marks on the carpet when the furniture is put back in place?

10. Explain why one caravan tilted when someone walked to one end of it.

11. Design and draw a bicycle stand which would be suitable for a heavily-loaded delivery bike.

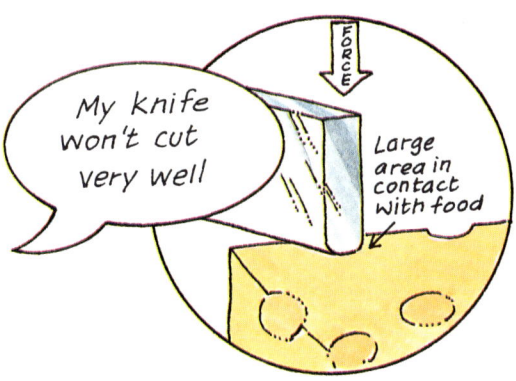

The area of contact of the knife edge will alter the pressure of the blade on the surface. For the same force, the blunt knife will have a lower pressure because its surface area is larger than that of the sharp knife.

12. Design a simple experiment which would allow you to estimate the surface area of both cutting edges.

13. Design a knife which would keep its cutting edge as sharp as possible—even though some of the edge is bound to wear away with use.

8

SIMPLE ELECTRICAL DEVICES

This is a device used to light a gas fire. The cross-sectional diagram shows the electrical circuit in the gas lighter. When the switch is pressed, electricity can flow from the battery through the wire at the end of the lighter. The wire gets hot and glows.

1. Here is a circuit diagram for the gas lighter:

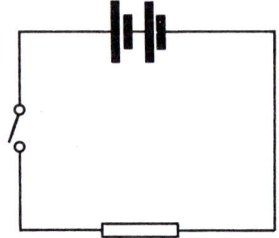

Identify the following symbols in the circuit diagram.

(a) ┤├ (b) ─o╲o─ (c) ─▭─

2. Energy transfer can be represented by arrow diagrams. This arrow diagram shows energy transfer in a gas lighter:

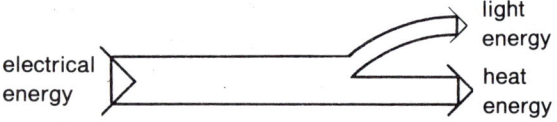

Why is it useful that a small amount of the electrical energy in a gas lighter is changed to light energy?

A battery-powered torch works in a similar way to the gas lighter. This time, when the switch is closed, the electricity flows through the bulb.

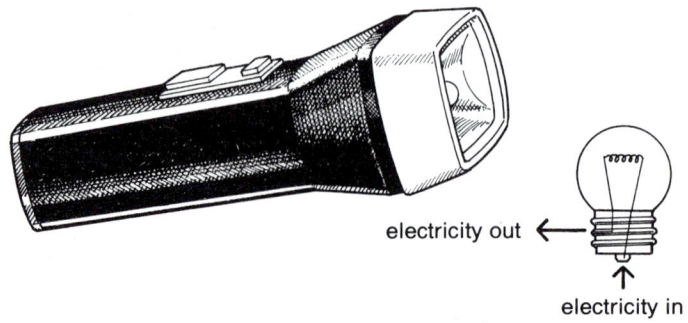

electricity out ←

↑
electricity in

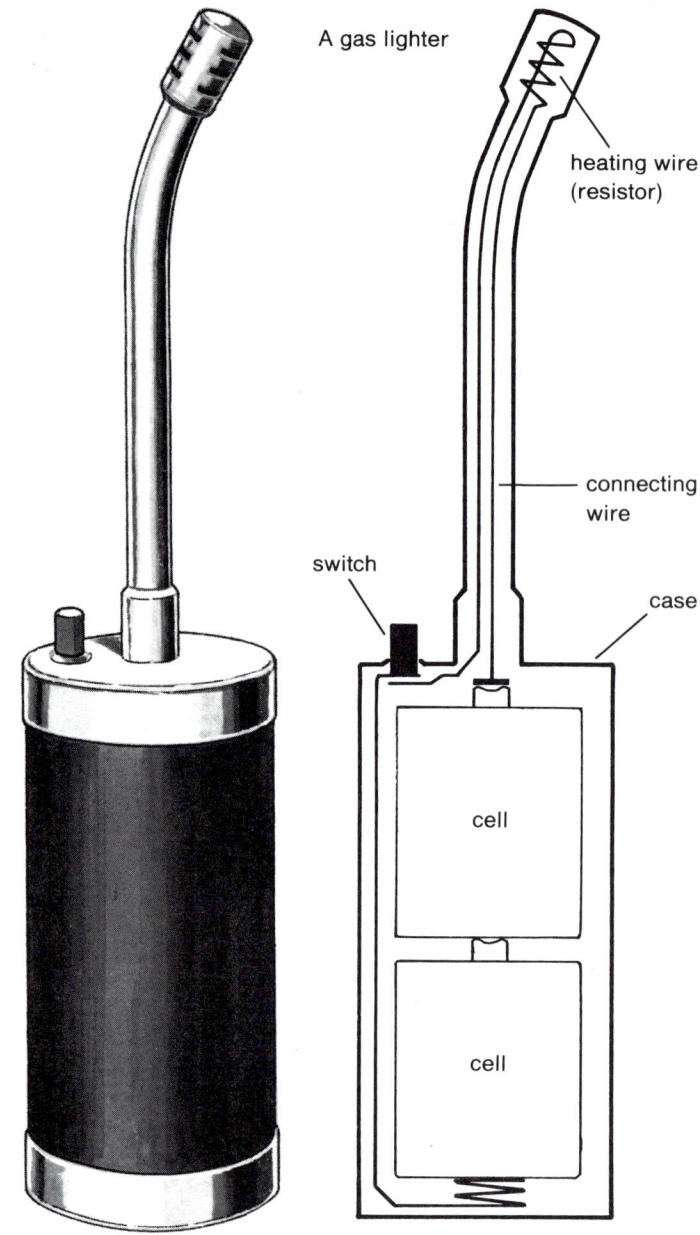

A gas lighter

heating wire (resistor)

connecting wire

switch

case

cell

cell

3. Draw a diagram to show how the parts of the torch pictured here are assembled to give a complete circuit.

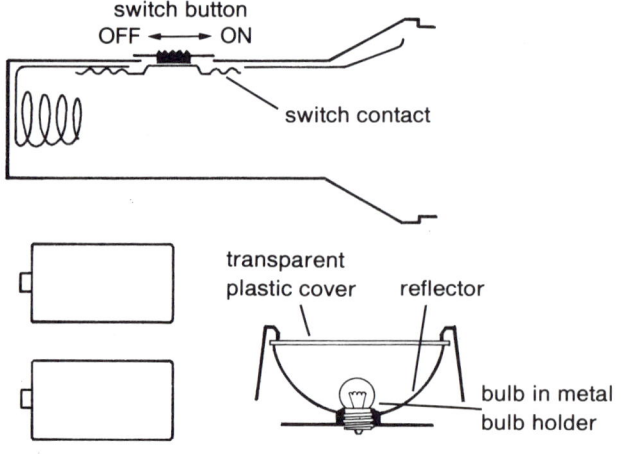

switch button
OFF ← → ON

switch contact

transparent plastic cover

reflector

bulb in metal bulb holder

4. Draw a circuit diagram for the torch. The symbol for a bulb is ─○─

5. Copy out and label the arrow diagram below to show the energy transfer for a torch.

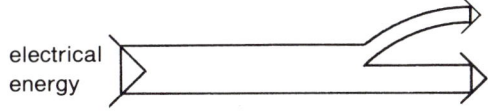

electrical energy

6. Suggest TWO differences between the wire used in the bulb and the wire used in the gas lighter which might explain why one produces more light than the other.

Many people use a battery-powered fan to keep cool. When the switch is closed, electricity passes through the motor. The motor turns the fan. Some of the electrical energy is transferred to heat energy as the motor is running.

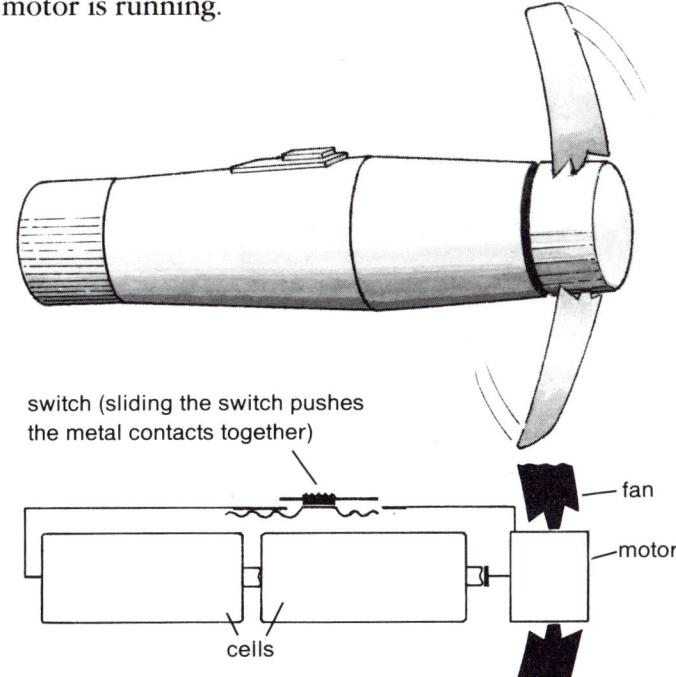

switch (sliding the switch pushes the metal contacts together)

fan

motor

cells

Cells for electrical devices come in various sizes. All of these cells are 1.5 volts. The more chemicals there are in a cell, the longer it will last.

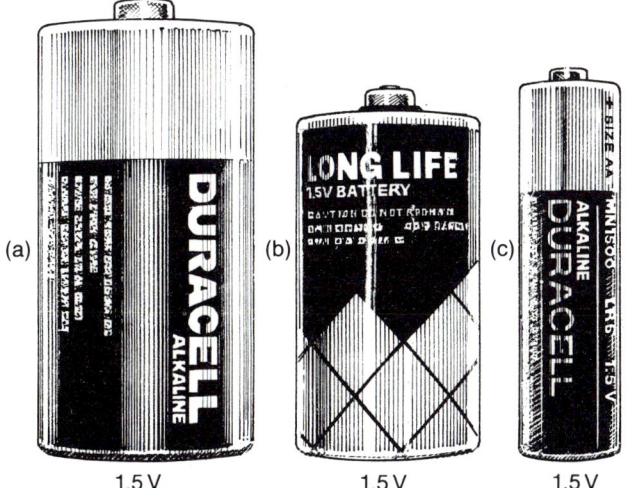

(a) 1.5 V   (b) 1.5 V   (c) 1.5 V

7. The symbol for a motor is ─○─. Draw a circuit diagram for the fan.

8. Copy out and complete the arrow diagram below to show the energy transfer for a fan.

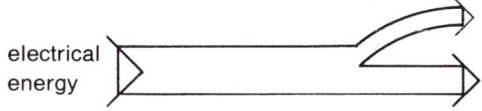

electrical energy

9. Use the facts you know about electrical devices to design a hair drier. Your design should include a fan and a heater. You can show your design as a drawing or as a circuit diagram.

10. Why might a battery-driven hair drier not be very effective?

11. Which cell, (a), (b) or (c) do you think would last the longest in a torch?

12. How many of these cells would be needed to run a 9 volt transistor radio or personal stereo?

13. Digital watches use very small cells called "button cells" in them. However, they can last a very long time. What does this tell you about the amount of electrical energy needed to run a digital watch, compared with a torch?

9

ELECTROSTATICS— FRIEND OR FOE?

BREAKTHROUGH IN PLANT SPRAYING

Engineers at the local firm of Stanley Bros. have developed a new device which they hope will capture a home market and bring badly needed orders from overseas.

The device, called 'Econospray', sprays pesticides on plants. It uses a simple idea in physics called electrostatics. Static electricity is familiar to us all when things get 'charged up' by friction. The electric charges on two objects make them repel each other if the charges are the same, or attract if they are opposite. Charged objects will also attract uncharged objects.

Savings and efficiency

Giving the droplets of pesticide an electrostatic charge as they come out of the nozzle makes them repel each other. The droplets come out in a finer spray, which covers the plant more evenly.

But the really big advantage is that the spray sticks to the plant. The droplets stick to the leaves in the same way that a party balloon sticks to a wall when you rub it, and a thin stream of water is attracted by a charged plastic ruler.

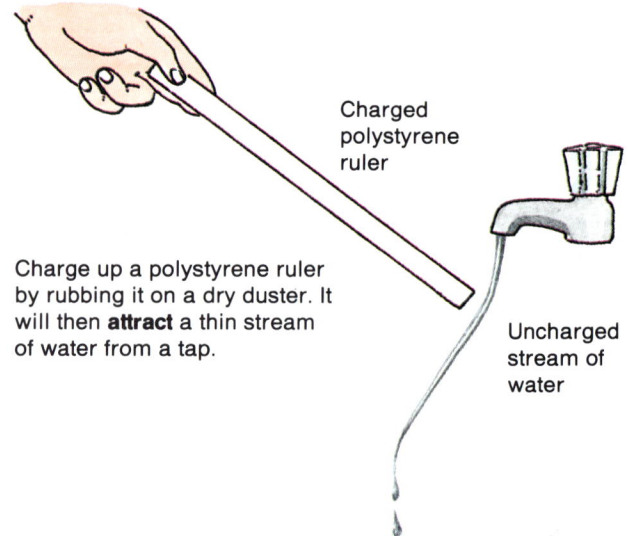

Charged polystyrene ruler

Charge up a polystyrene ruler by rubbing it on a dry duster. It will then **attract** a thin stream of water from a tap.

Uncharged stream of water

Wasteful

Very often, horticulturists with hundreds of plants in greenhouses use more chemical spray than they need to, just to make sure that plants get covered. Insects thrive in greenhouses where it is warm and there are no predators, so plants need a thorough coating of pesticide.

1. Imagine that you work for Stanley Bros. and you are in charge of marketing the Econospray. You have to design a leaflet which could be given to horticulturists to encourage them to buy the spray. Remember, horticulturists are busy people so the leaflet should be short. It should

answer the following questions:

● Why should I bother changing to a new spray?
● How does the spray work?
● How will it save money?

2. Look at the picture of paint being sprayed on to the shells in the ammunition factory. To get an even coating of paint a method similar to the one for plant spraying is used. A charged droplet of paint will only be attracted to the shells if the charge on the droplet is opposite to the shells or the shells are connected to earth (like the plant in the ground).

(a) What is done to the paint as it comes out of the jet?
(b) Where do you think that a wire leading from the shells would be connected to?

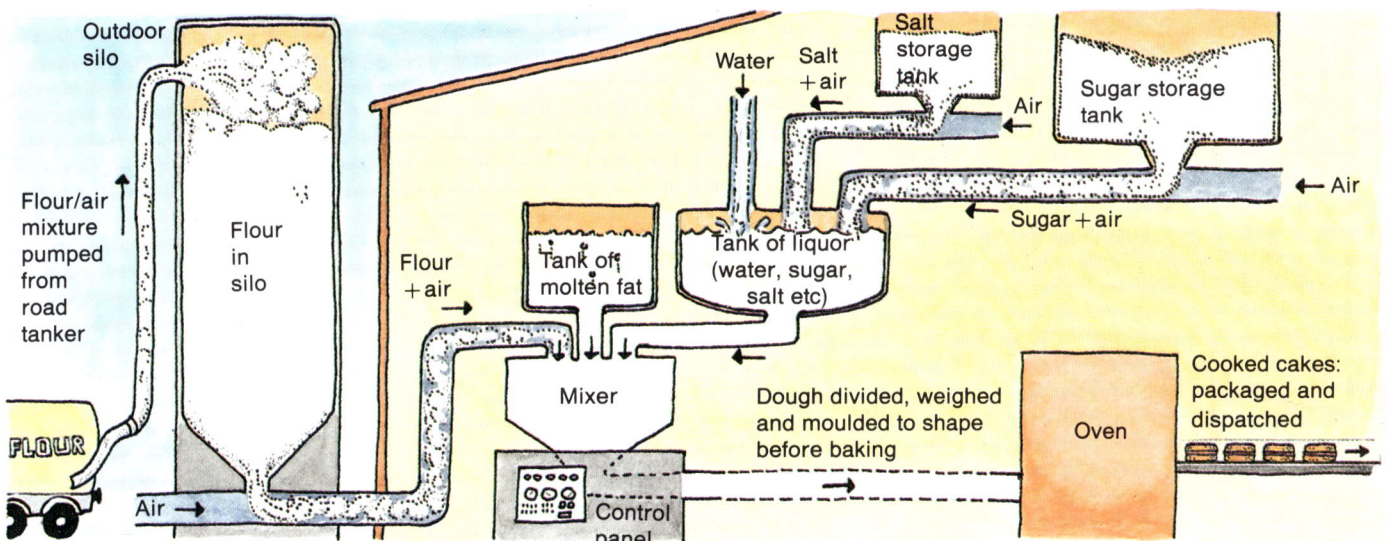

DIAGRAM OF THE BAKERY

Explosion at local bakery

Last Wednesday there was a small explosion and fire at the Mill Bakery in Station Road. Fire engines were called in and soon stopped the fire. Fortunately, no-one was hurt. The owners have started an enquiry to find the cause of the explosion.

The firm uses large amounts of powders such as sugar and flour. When sugar and flour are in powder form their surface area is increased. This means that they will flow easily, but it also means that they will burn more easily. The powders are stored in large tanks above the departments where they are going to be used, or in outdoor silos. Then the powders flow through pipes to the various departments where they are needed. When the powders rub against the pipes they can 'charge up'. Powder particles mixed with air are more likely to charge up as they swirl around, rubbing against each other and the pipe. If someone touches a pipe, a spark can jump to the person on its way to "earth". An artificial earth should be provided by a wire connected to the ground (like a lightning conductor) to conduct charges safely without sparks.

3. Sketch the diagram of the bakery. Label THREE places where there might be a build-up of electrostatic charge. Number them 1, 2 and 3.

4. Imagine that you are a safety expert who has been called in to investigate the fire at the bakery. Using the information on this page, write a letter to the firm. You should:

(a) explain why powdered sugar and flour burn easily.
(b) explain how you think the fire started.
(c) suggest what might be done to improve safety.

10

ELECTROMAGNETS IN BUILDINGS

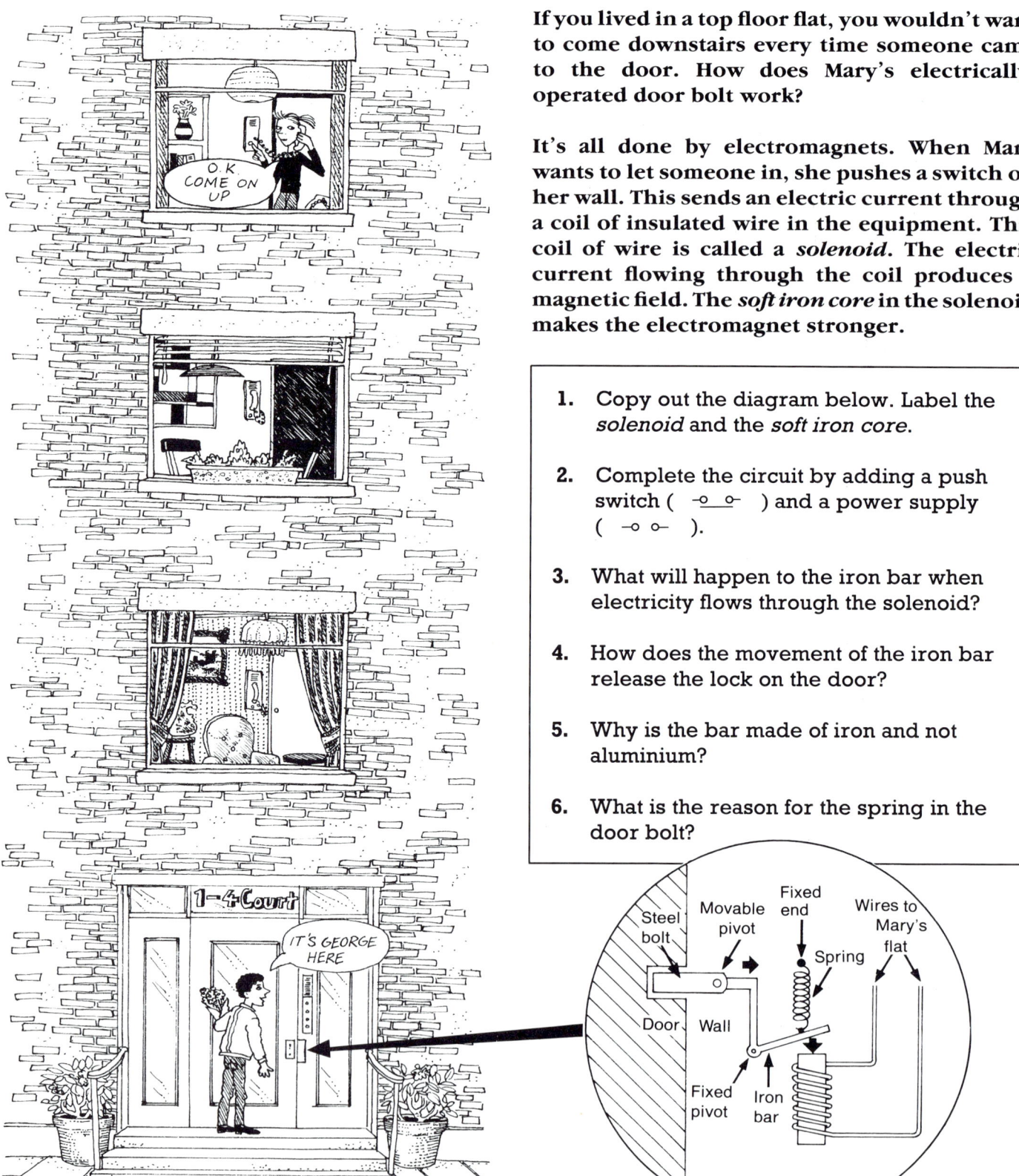

If you lived in a top floor flat, you wouldn't want to come downstairs every time someone came to the door. How does Mary's electrically-operated door bolt work?

It's all done by electromagnets. When Mary wants to let someone in, she pushes a switch on her wall. This sends an electric current through a coil of insulated wire in the equipment. This coil of wire is called a *solenoid*. The electric current flowing through the coil produces a magnetic field. The *soft iron core* in the solenoid makes the electromagnet stronger.

1. Copy out the diagram below. Label the *solenoid* and the *soft iron core*.

2. Complete the circuit by adding a push switch (—o͏ o—) and a power supply (—o o—).

3. What will happen to the iron bar when electricity flows through the solenoid?

4. How does the movement of the iron bar release the lock on the door?

5. Why is the bar made of iron and not aluminium?

6. What is the reason for the spring in the door bolt?

This idea can be used to make another kind of **electromagnetic switch**. Sometimes it is necessary to switch very large electric currents on and off. This could be dangerous. For example, very powerful electric motors are needed to move a lift up and down a building. These motors use large currents of electricity. The wires carrying these currents should not directly connect with the circuit of the switch that people in the lift have to press. If there were a fault and you pressed the switch … you can imagine!!!

The solution is to use a **relay** switch. This switch works very much like the door bolt.

DIAGRAM OF A RELAY SWITCH

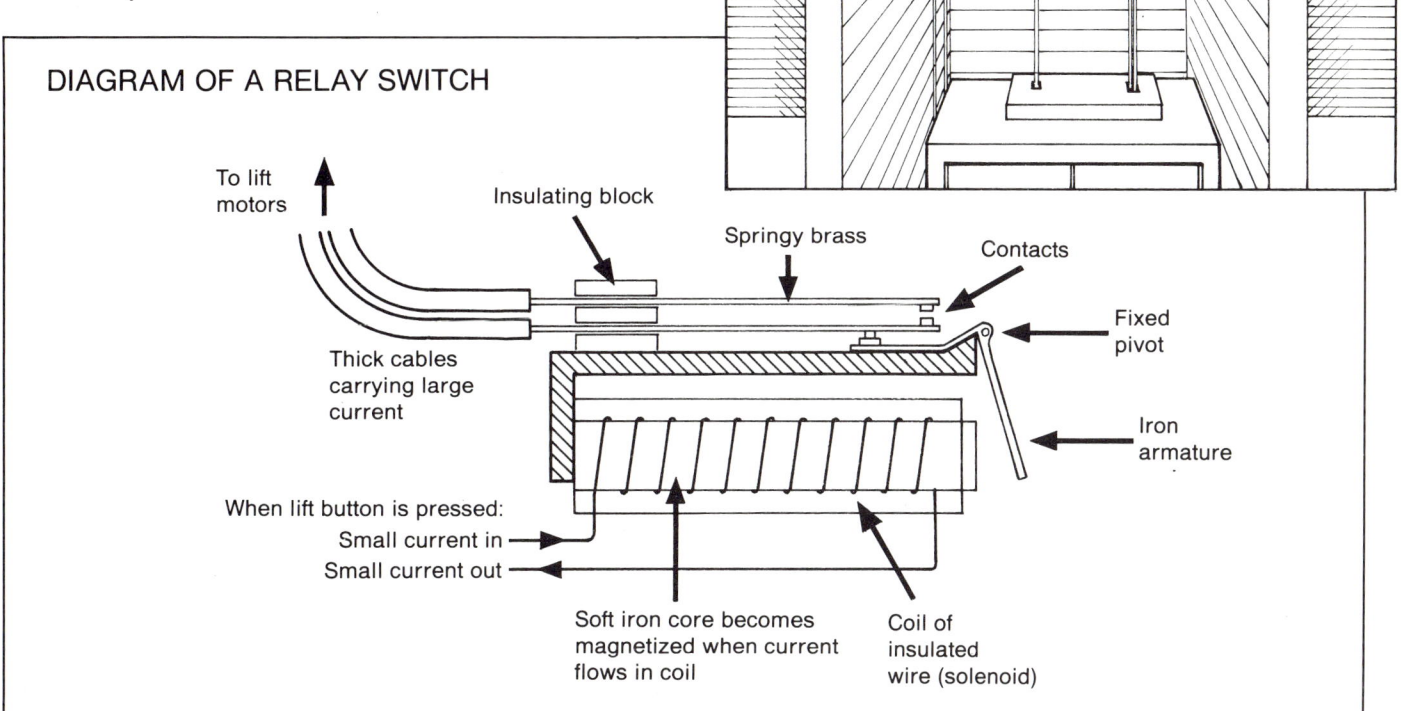

To lift motors

Insulating block

Springy brass

Contacts

Fixed pivot

Thick cables carrying large current

Iron armature

When lift button is pressed:
Small current in
Small current out

Soft iron core becomes magnetized when current flows in coil

Coil of insulated wire (solenoid)

---

7. What happens to the soft iron core when electricity flows through the solenoid?

8. What effect does this have on the iron armature?

9. Explain how the relay completes the circuit to switch on the lift motors.

10. How do you think the relay switches the lift motors off?

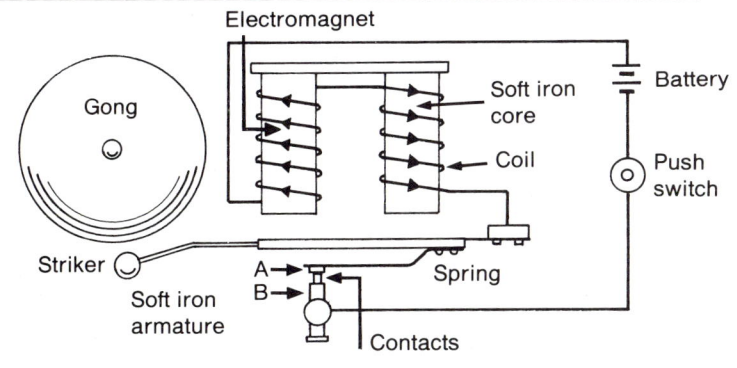

Electromagnet

Gong

Soft iron core

Battery

Coil

Push switch

Striker

Soft iron armature

A
B

Spring

Contacts

---

11. Electric doorbells use electromagnets as well. Use the diagram to complete the sentences below. Copy out each sentence and fill in the missing words.

(a) When the push switch is pressed a current flows in the c___ of the el_____.

(b) The s___ i___ a_____ is attracted by the e_____.

(c) This causes the s_____ to hit the g___.

(d) The movement of the armature separates the c_____ at__ and __.

(e) Electricity stops flowing in the c___ because the circuit is now broken.

(f) The e_____ is no longer magnetized, so the a_____ moves back again, pulled by the s_____.

LOGI-COOKIES

THIS PART OF THE MACHINE MIXES THE SUGAR, FLOUR, BUTTER AND THE EGGS.

In the Logi-Cookie bakery, there is a large automatic machine which mixes all the ingredients together and bakes the cookies.

Logic gates have been used to make sure the mixers do not switch on unless all the correct ingredients are present.

There are sensors to detect the ingredients entering the machine. These are connected to AND and OR gates. If the signal going to a mixer is a ONE, the mixer will switch on, if it is a ZERO, it will not switch on.

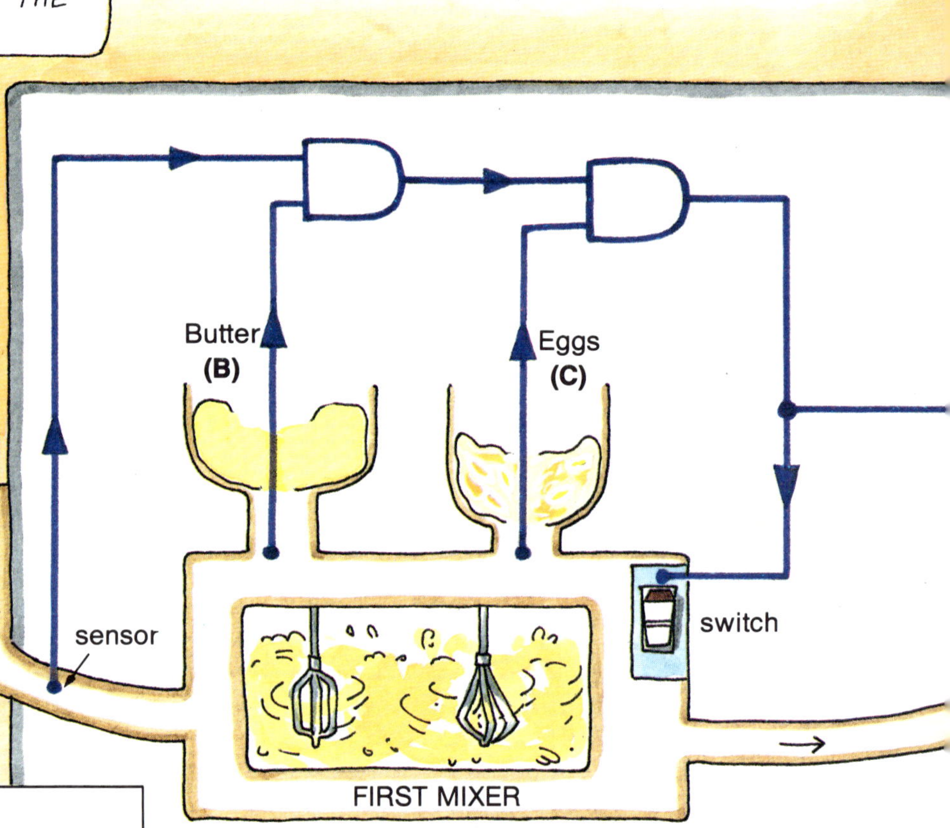

Butter **(B)**

Eggs **(C)**

Flour and sugar **(A)**

sensor

switch

FIRST MIXER

1. What type of gate is this?

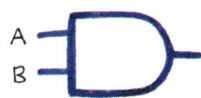

The first AND gate is connected to the flour and sugar sensor A and the butter sensor B. This truth table shows what happens to the output (1 means ingredient present, 0 means no ingredient). There are four possibilities:

Flour and sugar (A)	Butter (B)	Output of AND gate
0	0	0
0	1	0
1	0	0
1	1	1

With three ingredients, A, B and C, there will be eight possibilities.

2. Copy and complete this table to show which ingredients switch the mixer on.

Flour and sugar (A)	Butter (B)	Eggs (C)	1st AND gate	2nd AND gate	Mixer ON/OFF
0	0	0	0	0	OFF

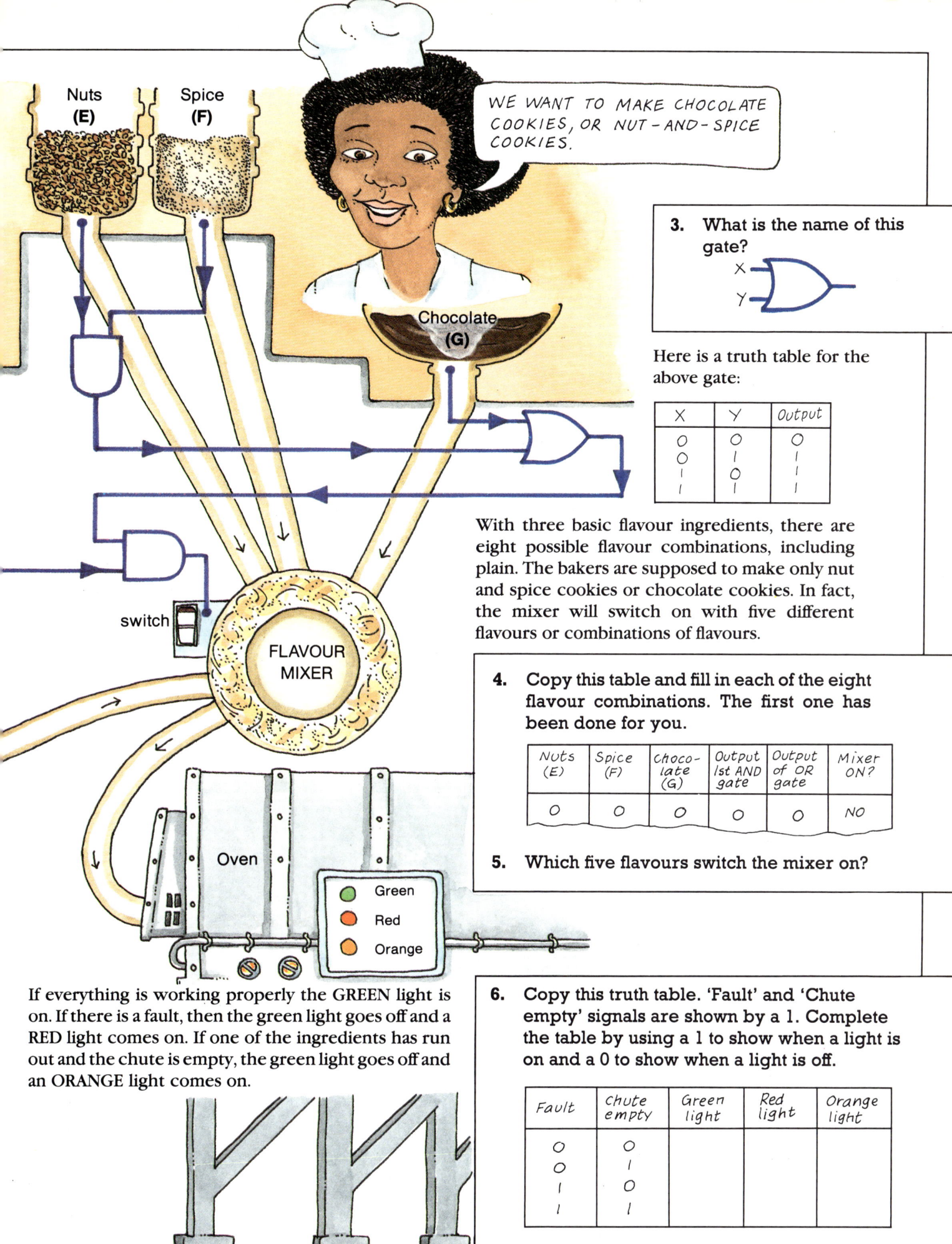

Nuts
(E)

Spice
(F)

WE WANT TO MAKE CHOCOLATE COOKIES, OR NUT-AND-SPICE COOKIES.

Chocolate
(G)

switch

FLAVOUR MIXER

Oven

- 🟢 Green
- 🔴 Red
- 🟠 Orange

If everything is working properly the GREEN light is on. If there is a fault, then the green light goes off and a RED light comes on. If one of the ingredients has run out and the chute is empty, the green light goes off and an ORANGE light comes on.

3. What is the name of this gate?

Here is a truth table for the above gate:

X	Y	Output
O	O	O
O	I	I
I	O	I
I	I	I

With three basic flavour ingredients, there are eight possible flavour combinations, including plain. The bakers are supposed to make only nut and spice cookies or chocolate cookies. In fact, the mixer will switch on with five different flavours or combinations of flavours.

4. Copy this table and fill in each of the eight flavour combinations. The first one has been done for you.

Nuts (E)	Spice (F)	Choco-late (G)	Output 1st AND gate	Output of OR gate	Mixer ON?
O	O	O	O	O	NO

5. Which five flavours switch the mixer on?

6. Copy this truth table. 'Fault' and 'Chute empty' signals are shown by a 1. Complete the table by using a 1 to show when a light is on and a 0 to show when a light is off.

Fault	Chute empty	Green light	Red light	Orange light
O	O			
O	I			
I	O			
I	I			

SIMPLE

1. A **lever** has a pivot, somewhere to pull or push and some sort of load to move. Find THREE examples of levers in the picture and write them down.

2. An **inclined plane** is a slope. This can make jobs easier to do. Find THREE examples of inclined planes in the picture and write them down.

3. A **wheel and axle** helps us to move loads along more easily. Find THREE examples of a wheel and axle in the picture and write them down.

4. A **pulley** has a moving wheel with a groove in it and a rope or chain running in the groove. Find TWO examples of pulleys and write them down.

5. The thread of a **screw** has a spiral shape. Find THREE examples of screws and write them down.

6. A **wedge** has a slope like an inclined plane, but can also be a cutting edge. Like all simple machines, it helps us to do jobs more easily. Find THREE examples of wedges.

7. Say where you might find:

(a) a lever on a bicycle

(b) an inclined plane in your home

(c) a wheel and axle on a machine used for sport

(d) a pulley in a workshop

(e) a screw in the kitchen

(f) a wedge in the kitchen

Now explain what is being done by each of these machines.

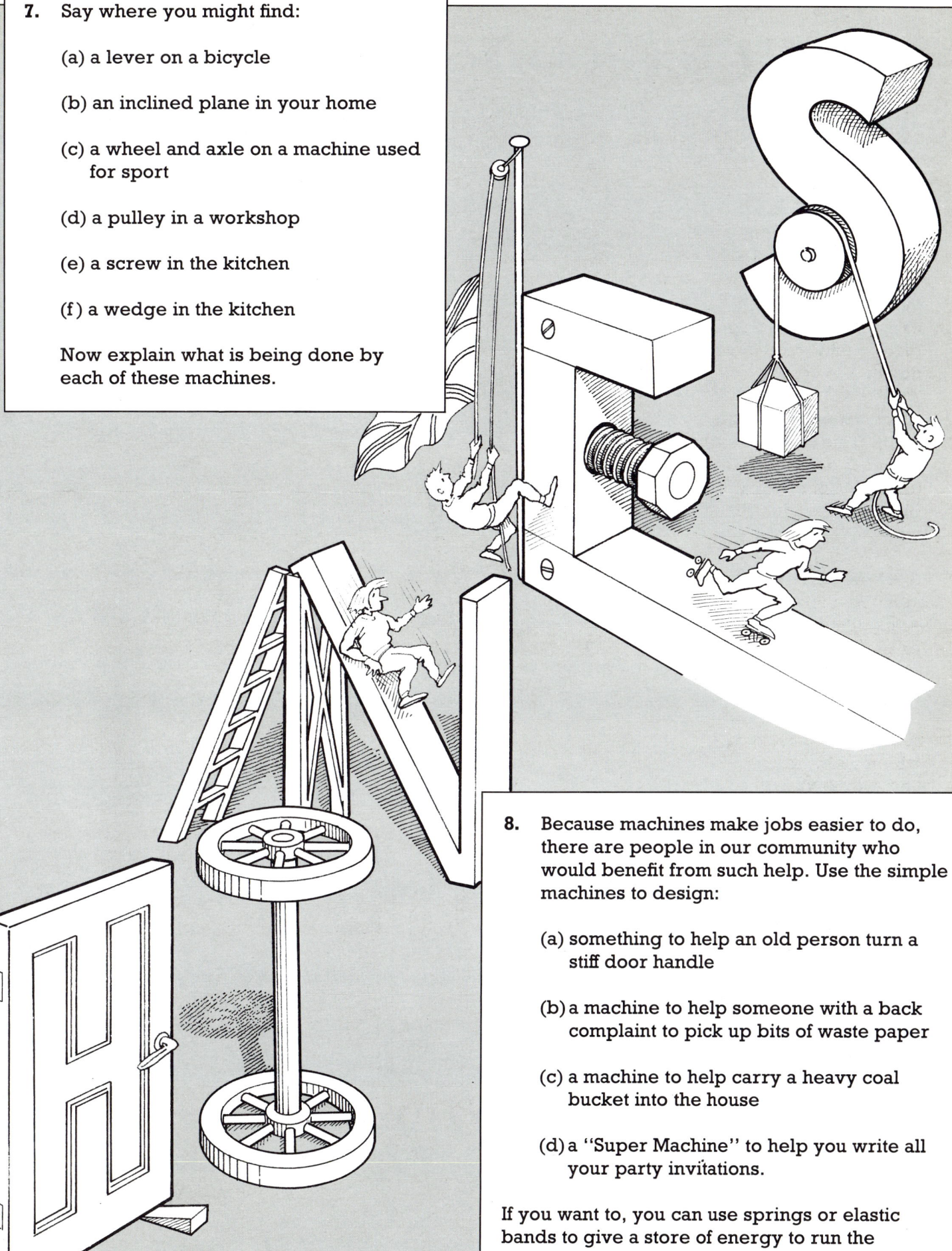

8. Because machines make jobs easier to do, there are people in our community who would benefit from such help. Use the simple machines to design:

(a) something to help an old person turn a stiff door handle

(b) a machine to help someone with a back complaint to pick up bits of waste paper

(c) a machine to help carry a heavy coal bucket into the house

(d) a "Super Machine" to help you write all your party invitations.

If you want to, you can use springs or elastic bands to give a store of energy to run the machines.

BICYCLES—FRICTION AND STABILITY

When you ride a bicycle, you do more than just push the pedals round. There are many parts of your body working. Think about the muscles you use when riding uphill—these are producing forces. The forces can involve pushes, pulls, stretching and twisting. Your weight is a force caused by gravity pulling your mass downwards. Friction is a force which is in the opposite direction to the force pushing forwards.

1. Suggest THREE places on the bicycle where there is friction that is helpful to a cyclist.

2. Suggest ONE place on the bicycle where there is friction that is unhelpful to a cyclist.

3. Draw a simple diagram of the bicycle and label the forces produced when you ride. Use arrows to show the direction of the forces. You should be able to think of four forces.

'Stable' means that an object will not tip over easily. Forces are balanced.

'Stable' objects are usually near the ground and have wide bases.

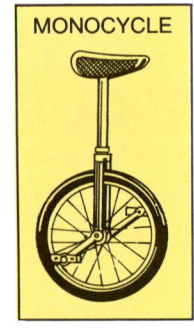

MONOCYCLE

CHILD'S BICYCLE WITH STABILISERS

PENNY FARTHING

ADULT BICYCLE No 1

ADULT TRICYCLE

Forces to drive the bicycle forward are produced when the pedals turn the wheels.

CHILD'S TRICYCLE

TANDEM

A **small wheel** goes round more times than a large wheel for the same distance travelled.

4. Which TWO bicycles will be the most stable? Explain why.

5. Which TWO bicycles will be the most unstable? Explain why.

6. Which bicycle can produce the most force to drive it forwards?

7. Which TWO bicycles would probably have their tyres worn out most quickly? Give a reason for your answer.

8. Which of the two adult bicycles would be the most stable? Explain how you decided.

9. Why is a stable bicycle important for road safety?

A load being carried on a bicycle produces a force. If the bicycle is to be stable, there must be 'balanced forces' to keep it upright.

10. Draw a picture of a bicycle and show the best position for carrying a large school bag. Write a few sentences to explain why you think this is the best position. Use the idea of 'balanced forces' to explain your choice.

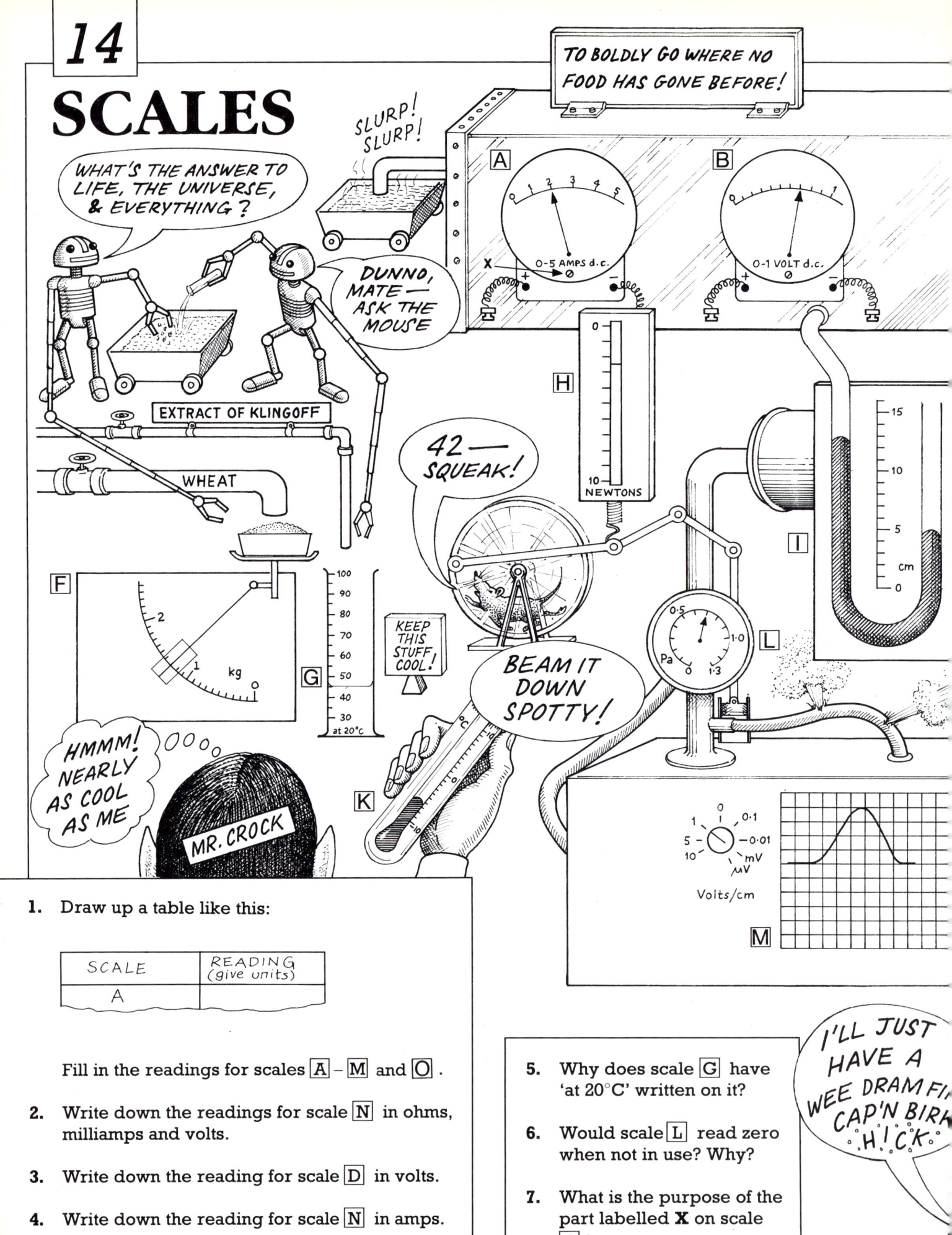

14 SCALES

1. Draw up a table like this:

SCALE	READING (give units)
A	

Fill in the readings for scales A – M and O .

2. Write down the readings for scale N in ohms, milliamps and volts.

3. Write down the reading for scale D in volts.

4. Write down the reading for scale N in amps.

5. Why does scale G have 'at 20°C' written on it?

6. Would scale L read zero when not in use? Why?

7. What is the purpose of the part labelled **X** on scale A ?

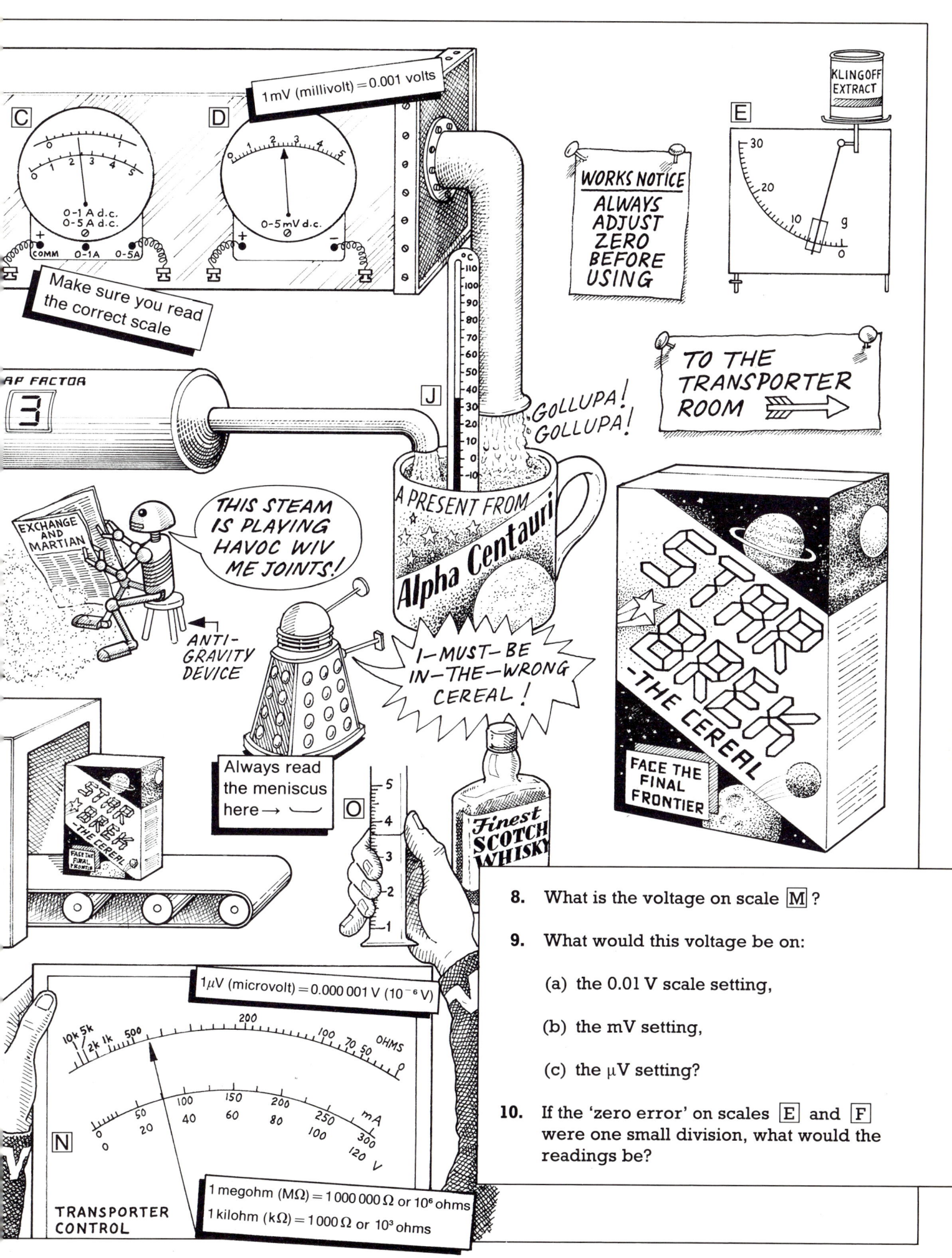

8. What is the voltage on scale $\boxed{\text{M}}$?

9. What would this voltage be on:

   (a) the 0.01 V scale setting,

   (b) the mV setting,

   (c) the μV setting?

10. If the 'zero error' on scales $\boxed{\text{E}}$ and $\boxed{\text{F}}$ were one small division, what would the readings be?

15 ENERGY RESOURCES

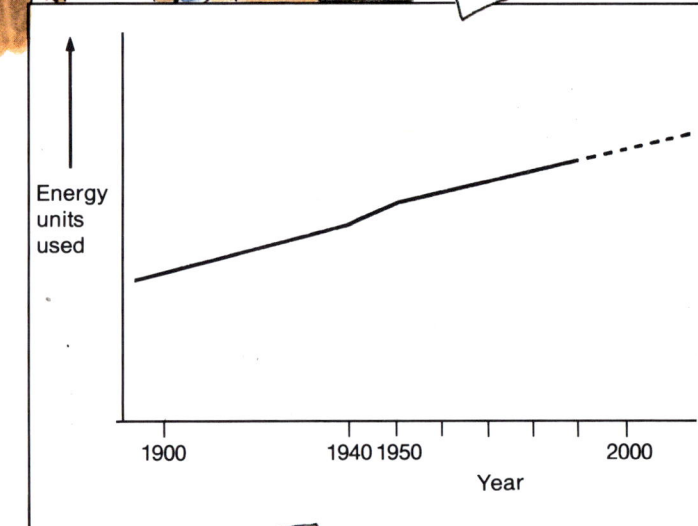

WORLD ENERGY BANK

Ladies and gentlemen, you are the leaders of the countries of the world. I have called you together to discuss your energy account at this—The World Energy Bank.

The situation is serious. Last year you used even more energy—making it another record spending spree!

Here is a graph to show you how your spending has increased over the years since 1900.

1. Why do you think each of the following things have meant an increase in the use of energy?

(a) the increase in population
(b) changes in transport systems
(c) increases in wealth

2. Suggest THREE things we could do in this country to cut down our energy spending.

3. What happened to the number of energy units used between 1940 and 1950? Suggest a reason.

THE ENERGY BANK plc

Customer: The Earth

DEPOSIT ACCOUNT : Non-Renewable Resources

	ENERGY UNITS
COAL ⎫	210 000
OIL ⎬ Fossil Fuels	32 000
GAS ⎭	22 000
NUCLEAR	950 000

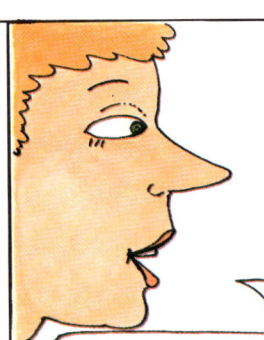

Remember that the coal, oil and gas took a very long time to form in the Earth. Once they are used up they will be very difficult to replace. This is why they are in your "Non-renewable Resources" account. The figures for nuclear fuels may look good, but even these are non-renewable—once used, they cannot be replaced. And remember, nuclear fuels may cause problems because some of the waste from reactors stays radio-active for a long time.

4. What is a "non-renewable resource"?

5. Which fossil fuel has:
(a) the largest reserves?
(b) the smallest reserves?

6. What is the total supply of fossil fuels available to us?

But what about our income? We've got new energy coming in all the time!

You mean renewable resources. Yes, you have. For example, a lot of sunlight falls on the Earth each year—far more energy units than you need. The trouble is, it's too spread out. True, it gets trapped and concentrated in plants like sugar-beet and wood as a result of photosynthesis. But your income from these biomass fuels is still quite small.

Nuclear reactors which breed their own fuel look promising, but there are people who think that these may cause a health or pollution problem for the future.

THE ENERGY BANK plc
Customer: The Earth
CURRENT ACCOUNT: Renewable Resources

	ENERGY UNITS
SUN ON LAND	2 800 000
CROPS	3
WIND	3
GEOTHERMAL	2
TIDAL	20
HYDRO-ELECTRIC	300
FARM WASTE	200
BURNING WOOD	4
NUCLEAR BREEDER REACTOR	95 000 000

7. Name the TWO biomass fuels mentioned in the text.

8. Name the process in which plants store the sun's energy in chemicals such as sugar and starch.

9. What is the connection between the sun's energy and the energy of:

   (a) the wind?
   (b) the energy stored in the water of reservoirs?

The best estimate which I have for your future energy bills is about 47 000 energy units by the year 2000. But don't forget—you will be getting this bill, plus the interest, every year after that. This bill of 47 000 units is about twice what you were spending in 1960!

So what do we do?

Ladies and gentlemen, I can't tell you how to run your world. One thing is clear: if you don't develop more renewable resources in the future you may end up bankrupt!

10. Approximately how long would fossil fuels last **after** the year 2000—bearing in mind the estimates for the energy bill in that year?

16
THE NUISANCE OF NOISE

Noise is loud, unwelcome sound. It can be a big problem in industry. Machines often make too much noise.

Loudness is measured in decibels (dB). This scale shows the loudness of some different sounds.

Decibels (dB)

Decibels	
150	Immediate damage to ear
140	Pain begins
130	
120	Road drill or power saw (1 m away)
110	Paint sprayer
100	Mower or punch press
90	Large electric motor
80	Lathe
70	Vacuum cleaner
60	Typewriter
50	
40	Normal conversation
30	Rustling of leaves
20	
10	
0	Threshold of hearing

1. What is the loudness in decibels of
   (a) a lathe?   (b) a punch press?

2. Look at the above picture. Estimate the noise level the worker not wearing ear muffs is exposed to.

3. Why might a high noise level in a factory cause an accident?

4. Estimate the noise level in the room where you are now. What is causing this noise?

This table shows the percentage of people of different ages who have hearing loss because they have worked for long periods in noisy factories.

Exposure level (dB)	Age (years)			
	18	28	38	48
90	0	7	12	16
100	0	21	36	41
110	0	46	68	74

5. What percentage of people suffer from hearing loss by the age of 28 as a result of working in places where the noise level is
   (a) 90 dB?   (b) 110 dB?

6. Copy out and complete the following sentence. "The greater the noise level a worker is exposed to, the ..."

7. What percentage of people working in places where the noise level is 100 dB suffer from hearing loss by the age of
   (a) 28?   (b) 48?

8. Copy out and complete the following sentence. "The longer a person works in a noisy place, the ..."

Some governments have set time limits for people working in places where the noise reaches certain levels.

Maximum noise exposure times in

(a) USA

Sound level (dB)	90	92	95	97	100	102
Maximum exposure time (hours per day)	8	6	4	3	2	1

(b) UK

Sound level (dB)	90	93	96	99	102	105
Maximum exposure time (hours per day)	8	4	2	1	$\frac{1}{2}$	$\frac{1}{4}$

9. Which country, the USA or the UK, is trying to set the stricter standards for factory workers?

10. Suggest a reason why one country might set stricter noise control standards than another.

Sound is a wave motion caused in the air when a vibrating object sets the nearest particles of air in motion. This motion rapidly spreads out and is picked up by the ear. The outer ear on the side of the head collects the sound and channels it down the ear canal to the ear drum. The ear drum is like the skin on a tiny drum. As it vibrates, it sets the three tiny bones in the middle ear in motion. This motion passes to a coiled tube called the cochlea. Inside the cochlea are sensory cells which detect the movement and send messages to the brain. These cells can become damaged if a person is exposed to loud noise.

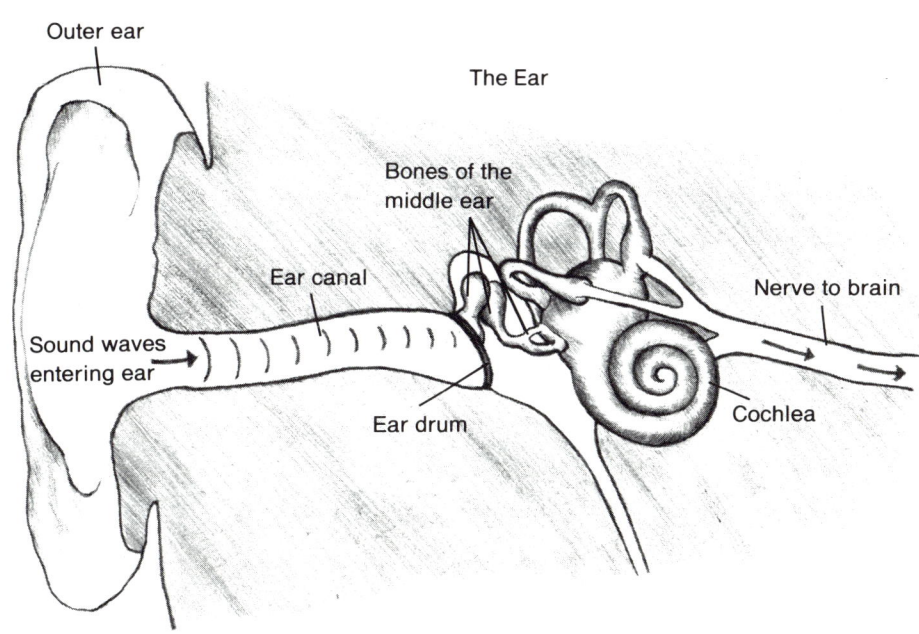

The Ear

Many people wear ear protectors when working in noisy places.

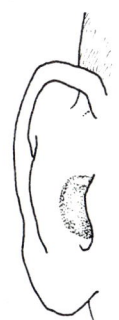

(a) Cotton wool plug

Reduces noise level by up to 10 dB. Can work loose and has to be pushed back.

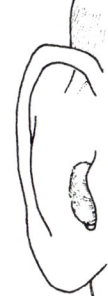

(b) Greased cotton wool plug

Reduces noise level by up to 30 dB. Can work loose (unhygienic if pushed back with dirty hands).

(c) Plastic ear plug

Reduces noise level by up to 30 dB. Fits tightly. Can be uncomfortable and cause sweating. Needs to be kept clean.

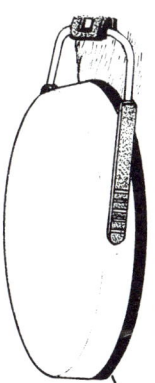

(d) Ear muffs

Plastic shell envelops ear. Contains sound-absorbing material. Reduces sound level by 50 dB. Can be uncomfortable and cause sweating.

12. Which type of ear protection would you recommend to someone using a power saw for short periods of time? Why?

13. Design a poster which could be placed near a noisy machine to encourage workers to wear ear protection.

SOUND SYSTEMS

When a loudspeaker is working, the cone vibrates. In each vibration, as the cone moves outwards, it pushes air particles in front of it closer together (high pressure). As it moves inwards, it spreads the air particles apart (low pressure). When this vibrating air reaches our ear-drum, it moves it inwards and outwards.

The layers of compressed and spread out air are "sound waves". The graph shows the pressure changes through a sound wave. The height of the wave above the central line is the **amplitude**. All of the sound systems shown in this unit have **amplifiers,** which control how loud the sound is. Loud sounds have large amplitudes. The power of an amplifier is measured in watts (W).

The number of vibrations of a speaker cone in one second is called the **frequency**. If a speaker cone vibrates 1000 times every second, its frequency is "1000 per second" or "1000 hertz" (1000 Hz). This is usually written as 1 kilohertz or 1 kHz.

Different speakers are used in sound systems to produce different ranges of sound. Large speakers are called **woofers**. They vibrate slowly and produce low-pitched (bass) sounds.

Very small speakers are called **tweeters**. They vibrate quickly and produce high-pitched (treble) sounds.

Mid-range speakers produce the sounds between low and high pitch.

A "three-way" speaker system has a woofer, a tweeter and a mid-range speaker. A stereo three-way speaker system has six speakers.

1. Look at the sound systems on the right. Which sound systems will be the loudest when the volume is turned up fully?

2. Use diagrams (a) and (b) to explain the difference between a loud and a quiet sound.

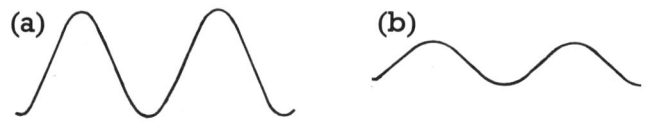

3. What is the connection between frequency and pitch?

4. Which TWO of the five sound systems best provide the full range of frequencies?

When a speaker vibrates at a high frequency, many waves are produced each second. A low frequency, and so a low pitch (or base sound) produces fewer waves per second.

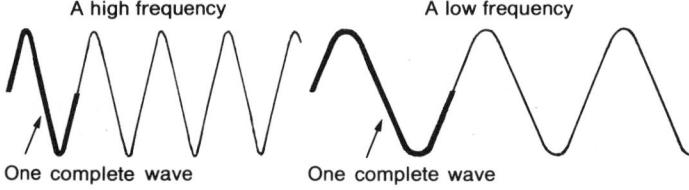

5. The diagrams show the number of complete waves in a certain time from some speakers.

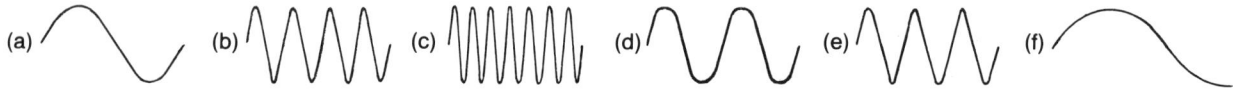

(a) Write down the number of complete waves for each of the diagrams (a)–(f).

(b) Which TWO of these sound waves would be produced by
(i) a woofer?   (ii) a tweeter?   (iii) a mid-range speaker?

£129.99

Detachable two-way, four speaker system. 60 W amplifier.

£39.99

Two-way, four speaker system. 16 W amplifier.

£99.99

One-way, two speaker system. 20 W amplifier.

£199.99

Detachable three-way, six speaker system. 60 W amplifier.

£99.99

3D SUPER WOOFER

Three-way, six speaker system. 25 W amplifier.

Graphic equalisers allow you to control the frequencies of the music to suit your ear. By sliding a control, you can boost or cut certain frequencies. The more graphic equalisers there are, the more control you have over the frequency range you want to hear.

TABLE 1   Key: $\sqrt{}$ = the system has these facilities

SOUND SYSTEM		1	2	3	4	5
TWIN CASSETTE		√		√	√	
HIGH SPEED DUBBING		√		√	√	
AUTO-REVERSE				√	√	√
AMPLIFIER OUTPUT (W)		60	16	20	60	25
GRAPHIC EQUALISER		5		3	5	3
APPROX. SIZES (cm)	HEIGHT	22	20	14	23	16
	WIDTH	57	46	50	63	48
	DEPTH	19	13	18	18	14

Use the information next to each sound system and the information in Table 1 to answer these questions.

6. Which sound system, number 3 or number 4, gives you the most control over the frequencies you can hear? Why?

7. Which system costing less than £100 is the most portable? (HINT: Volume = H × W × D). What other piece of information do you need before you can be sure of your answer?

8. Sound systems 1 and 4 offer almost identical facilities, but the prices are very different. If you had the money to buy either of these systems, what "fair test" would you devise to choose between them?

SEEING WITH SOUND

1. Why do you think ultrasound is used in the above device, not audible sound?

2. Which of the following statements do you think explains why the device only has a range of about 9 m?

   A Objects further away than 9 m are not of such interest.

   B The device doesn't work properly.

   C Ultrasound gets absorbed and after 9 m the beam is too weak to return.

3. Why might a blind person prefer to use this device rather than having a guide dog or a white stick?

4. Why might a blind person prefer to keep a guide dog rather than use this new device?

"It's a new pop group."

"No, you're wrong."

"Then it's the latest craze in fashion sunglasses."

"Sorry, wrong again."

"Well, what **is** it?"

"It's people...
SEEING WITH SOUND!"

"What do you mean? Seeing with sound?"

"Let me explain. You may have heard of **ultrasonic** waves. These are sound waves with a high frequency. Humans can't hear them, but some animals, like bats, can."

"What do you mean, frequency?"

"Frequency is another word for pitch. You know, high-pitched sounds, and deep, low-pitched sounds. Now let me continue ... We all know that bats have poor sight. They use a kind of echo-sounding system to prevent them bumping into things."

"What have bats got to do with the sunglasses?"

"They're not sunglasses—Listen, this 'bat' idea is being used by scientists to help blind people 'see' where they are going. Ultrasonic waves are given out by a hand-held device and the echoes are picked up by an electronic receiver—sunglasses to you. The receiver turns the echo into a bleep in the headphones. A blind person can 'scan' the area in front to a range of about 9 metres. The faster the echoes return, the faster the bleeps get and the person knows an object is close."

"Clever stuff!"

Ultrasound is useful in other ways too. Captain Hall and her crew are on the treasure trail, thanks to ultrasound. They are following Old Bluebeard's map. They know that the sunken treasure ship is somewhere between Bawden Point and Devil's Spit. But where exactly is it?

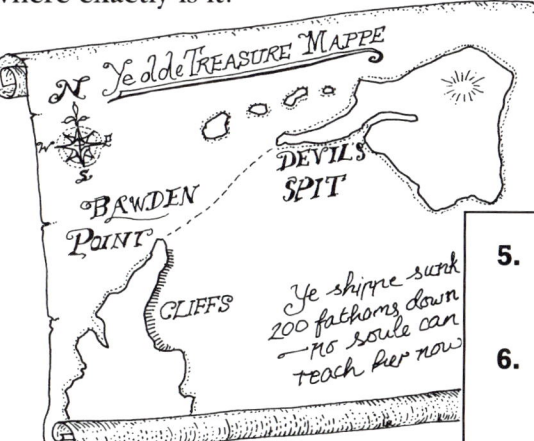

Captain Hall's ship has an echo-sounder. With this she can measure the depth of the sea and find sunken objects.

Ultrasonic waves are sent out from the ship and bounce off solid objects like the sea bed, submarines or sunken ships...

The echo is detected and the time taken for it to return is recorded. The shorter the time the echo takes to return, the shallower the sea.

Captain Hall's device produces ultrasonic waves that travel at 1430 metres per second in salt water.

5. How far does the ultrasound pulse travel through water in 6 seconds?

6. If the sea depth is 2000 m, how far does an ultrasound pulse travel from the time it is emitted to the time the echo is received?

Captain Hall piloted her ship from Bawden Point to Devil's Spit, stopping every 200 metres to make ultrasound measurements.

She could work out the depth of the sea by using the distance equation.

$$\text{DISTANCE} = \text{SPEED} \times \text{TIME}$$

travelled by ultrasonic wave   of ultrasonic wave   for echo to return

8. Draw a large graph of DISTANCE FROM BAWDEN POINT on the horizontal axis (0 to 4000 metres) against DEPTH OF SEA on the vertical axis (0 to 1500 metres).

9. Add a label **Y** to your graph to show where the sea is the deepest.

10. Add a label **X** to your graph to show where the sunken ship is. Use the graph to work out the depth of the sea at this point.

11. A fathom is a measure of distance.

One fathom = 1.85 metres

Use this information to find the depth of the sunken ship in fathoms.
Was Old Bluebeard right about the depth of his sunken ship?

12. One of the crew said Captain Hall was lucky to find the ship because 200 m is a large distance between measurements and the ship could easily have been missed. She said 50 m intervals should have been used. Captain Hall said there were a few reasons why she chose 200 m. Give one you can think of.

7. Use the equation to help you complete Captain Hall's results table.

Captain Hall's results table

DISTANCE FROM BAWDEN POINT (METRES)	TIME FOR ULTRASOUND WAVE TO GO DOWN AND BACK UP (SECONDS)	DISTANCE TRAVELLED BY THE ULTRASOUND WAVE (METRES)	DEPTH OF SEA (METRES)
200	—	—	—
400	0.1	140	70
600	0.3		
800	0.63		
1000	0.84		
1200	1.0		
1400	1.4		1000
1600	1.79	2560	
1800	1.79		
2000	1.96		
2200	1.79		
2400	1.6	2290	
2600	1.79		
2800	1.68		
3000	1.47		
3200	1.2	1720	
3400	1.0		715
3600	0.8	1140	
3800	0.67	960	
4000	—	—	—

19

HALL OF MIRRORS

These mirrors are not flat like the ones at home … That must be why they give such odd reflections.

A reflection of something in a mirror is called an **image**. Look carefully at the images in the mirrors, and at the three mirrors, **X**, **Y** and **Z**.

X

A plane mirror (flat)

A concave mirror

Y

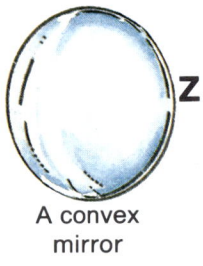

A convex mirror

Z

The light rays are being reflected from these two surfaces.

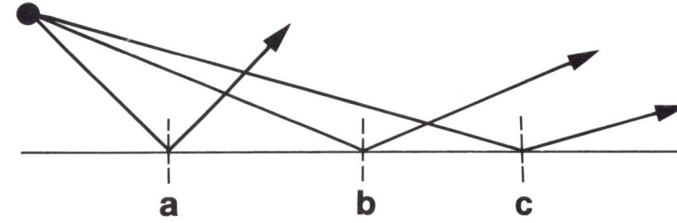

a　**b**　**c**

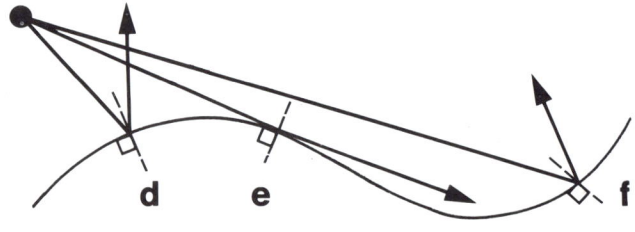

d　**e**　**f**

1. Which mirror (**X**, **Y** or **Z**) gives

   (a) a magnified (bigger) image?
   (b) a smaller image?
   (c) an actual size image?

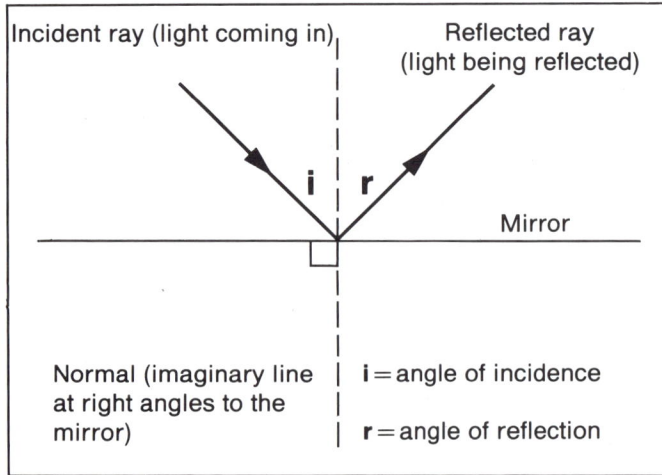

Incident ray (light coming in)

Reflected ray (light being reflected)

i　**r**

Mirror

Normal (imaginary line at right angles to the mirror)

i = angle of incidence

r = angle of reflection

2. Copy out this table.

	Angle of incidence (°)	Angle of reflection (°)
a		
b		
c		
d		
e		
f		

3. Use a protractor to measure the angles and fill in the values in your table.

4. What do you notice about the angle of incidence and the angle of reflection?

The sun is bright out here. It must have been quite dark inside the hall of mirrors.

All these pairs of sunglasses – and no two pairs are the same! I wonder which pair cuts out the most light?

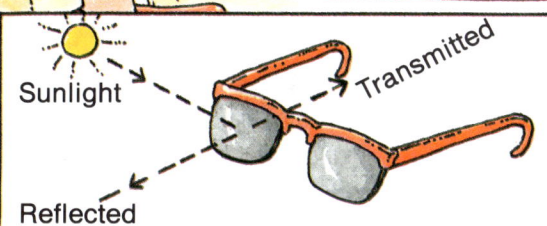

Sunlight

Transmitted

Reflected

The glasses which cut out the most light are the ones which transmit the smallest amount of light. Light which is not transmitted or reflected is absorbed by the lenses.

I bet you couldn't put all the sunglasses in order, starting with the darkest, and ending up with the least dark pair.

That would be hard. You would need a way of measuring the light transmitted through each pair.

Later . . .

I've thought of an experiment. To detect the amount of light we could use a solar calculator. The display switches off when there is not enough light.

Under a lamp, it needs 15 thin sheets of paper on top to switch it off.

With each pair of sunglasses we count the number of sheets of paper which are needed as well as the sunglasses to switch off the display.

RESULTS:

SUNGLASSES	SHEETS OF PAPER
YASMIN	7
PAUL	9
SUBHASH	6
GEMMA	14
DALEY	8
ASHLEY	7
KIM	12

5. Draw a bar chart of the number of sheets of paper needed for each pair of sunglasses.

6. Whose glasses cut out the most light?

7. Which two pairs of glasses were too similar to tell apart using this experiment?

8. Explain how you might be able to tell these two pairs apart.

9. Some glasses look like mirrors – what happens to most of the light which is not transmitted?

10. What happens to the light energy when light is absorbed by the lenses?

ANTIFREEZE

1 Dr Wilson, here is your new team member. I'll leave you to provide the background to the research project.

Dr Wilson is the project leader—you will be reporting to her.

2 Call me Louise – it's all first names here. It's good to have another scientist on the team.

We're aiming to produce an antifreeze for car radiators which can be used at very low temperatures, in countries around the Arctic for instance. Some of the big oil companies are interested.

5 Have you got the figures, Dave?

Here they are Louise. I've had a quick look at them and plotted a graph for R5. For X3, I think one of the readings is wrong—we'll need to check that, but there should be enough here to get the melting points.

Good, I've got to rush—a progress meeting starts in five minutes. You should find all you need on your desk. I'll come round and see how you're getting on before lunch.

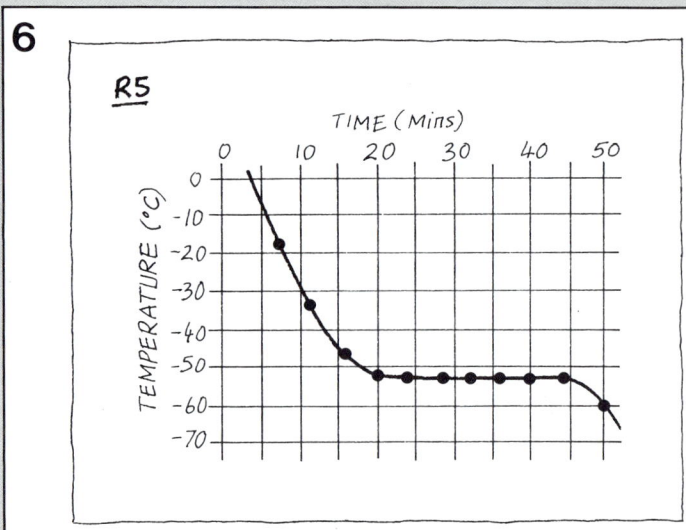

6 R5

TIME (mins) / TEMPERATURE (°C)

8 OK? Better get started!

1. What is the freezing point of R5?

2. Plot a graph of temperature against time for each of the other two liquids.

3. Which point on the X3 graph is likely to be wrong?

4. What is the freezing point of (a) TY6, (b) X3?

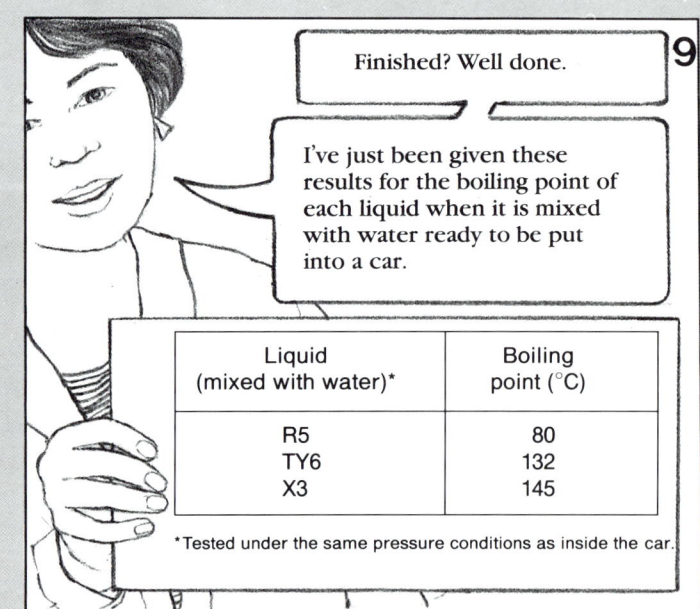

9 Finished? Well done.

I've just been given these results for the boiling point of each liquid when it is mixed with water ready to be put into a car.

Liquid (mixed with water)*	Boiling point (°C)
R5	80
TY6	132
X3	145

*Tested under the same pressure conditions as inside the car.

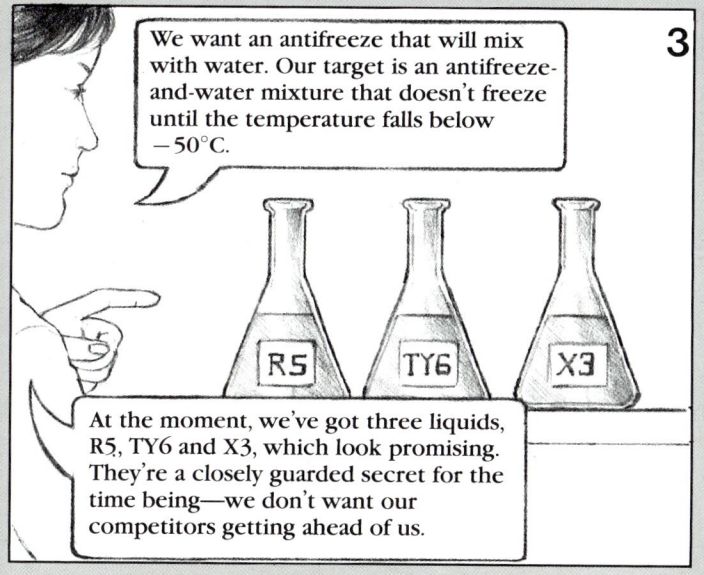

3 We want an antifreeze that will mix with water. Our target is an antifreeze-and-water mixture that doesn't freeze until the temperature falls below −50°C.

At the moment, we've got three liquids, R5, TY6 and X3, which look promising. They're a closely guarded secret for the time being—we don't want our competitors getting ahead of us.

4 This is Dave. He's done experiments to find out how the three pure liquids behave when they cool down in a special freezer.

Hi.

For your first job I'd like you to work with Dave on an analysis of his results. Then we need to see what happens when these pure liquids are mixed with water.

TY 6		X3	
Time (mins)	Temp (°C)	Time (mins)	Temp (°C)
0	5	0	0
4	−18	4	−25
8	−33	8	−43
12	−43	12	−53
16	−47	16	−62
20	−48	20	−65
24	−48	24	−67
28	−48	28	−65
32	−49	32	−68
36	−52	36	−68
40	−59	40	−69
44	−67	44	−71

7 You look a bit lost; let's make a list of the things to do.

Look at the graph of temperature against time for R5.

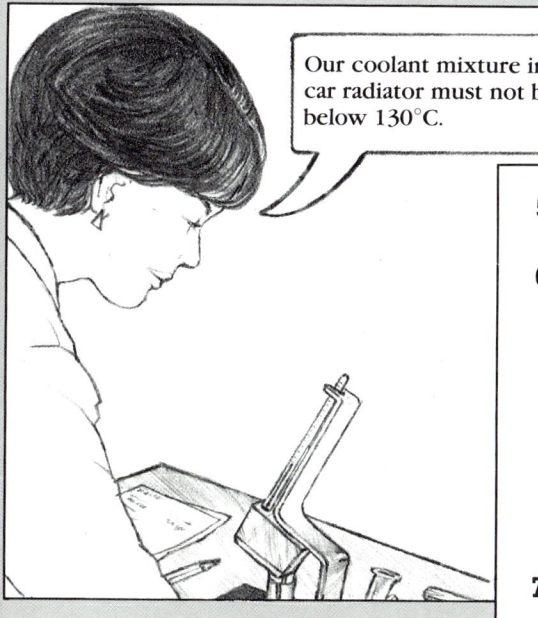

Our coolant mixture in the car radiator must not boil below 130°C.

10 Now your job is to find out when the antifreeze-and-water mixtures start to freeze. It won't be the same as the freezing point for the pure antifreeze liquids. We should be able to tell from a graph of temperature against time.

5. Which liquid or liquids would not boil below 130°C?

6. Design a method for finding the temperature at which antifreeze-and-water mixtures begin to freeze. The following points should guide your answer.
   ● What equipment will you use?
   ● What will you measure?
   ● What will you keep the same each time you do the experiment?
   ● How will you present your results?

7. Suggest TWO advantages of scientists working together on research projects in teams, rather than on their own.

Matrix of the Relationship between Activities and Attainment Targets of the National Curriculum

Activity key (columns): 1 Marathon · 2 Artificial Limbs · 3 Up, Up and Away · 4 Kettle · 5 On the Move · 6 Brake test · 7 Working under Pressure · 8 Simple Electrical Devices · 9 Electrostatics · 10 Electromagnets · 11 Logi-cookies · 12 Simple Machines · 13 Bicycles · 14 Scales · 15 Energy Resources · 16 The Nuisance of Noise · 17 Sound Systems · 18 Seeing with Sound · 19 Hall of Mirrors · 20 Antifreeze

Attainment Target	1	2	3	4	5	6	7	8	9	10	11	12	13	14	15	16	17	18	19	20
1 Exploration of science	✓	✓	✓	✓		✓	✓				✓									✓
6 Types and uses of materials	✓	✓		✓																✓
10 Forces		✓	✓		✓	✓	✓													
11 Electricity and magnetism								✓	✓	✓										
12 The scientific aspects of information technology including micro-electronics											✓									
13 Energy												✓	✓	✓	✓					
14 Sound and music																✓	✓	✓		
15 Using light and electro-magnetic radiation																			✓	

INDEX

Numbers refer to units, not pages.

Acknowledgements

The publishers would like to thank the following for supplying photographs for these units:

1 Rex Features

3 The Telegraph Colour Library

5 Camera Press (the space shuttle transporter); All-Sport (the Lotus); The Telegraph Colour Library (the Blackbird spy plane)

9 ICI Agrochemicals (the plant spray); The DeVilbiss Company Ltd (the paint spray)

13 Raleigh; Muddy Fox (the mountain bicycle); Bikerton Bicycles (the collapsible bicycle)

17 Toshiba Ltd; Hitachi Ltd; Sharp Electronics Ltd

Back cover: Cameron Balloons Ltd, Bristol